Robert Chapman

A Treatise on Ropemaking

As practised in Private and Public Ropeyard

.

Robert Chapman

A Treatise on Ropemaking
As practised in Private and Public Ropeyard

ISBN/EAN: 9783337350970

Printed in Europe, USA, Canada, Australia, Japan

Cover: Foto ©berggeist007 / pixelio.de

More available books at **www.hansebooks.com**

A TREATISE

ON

R.O P.E.M.A.K.I.N.G,

AS PRACTISED IN

PRIVATE AND PUBLIC ROPEYARDS,

DESCRIPTION OF THE MANUFACTURE

RULES, TABLES OF WEIGHTS,

ETC.,

ADAPTED TO

THE TRADE, SHIPPING, MINING, RAILWAYS, BUILDERS, &c.

BY

ROBERT CHAPMAN,

Formerly Foreman to Messrs. Huddart and Co., Limehouse; and Master
Ropemaker of H.M. Dockyard, Deptford.

Revised Edition.

LONDON:

E. & F. N. SPON, 48, CHARING CROSS.

1868.

LONDON:
Printed by W. CLOWES & SONS, Stamford Street and Charing Cross.

CONTENTS.

———◇———

B 2

INTRODUCTION.

THE Art of Ropemaking, by some strange fatality, has not attracted hitherto sufficiently the notice or attention of the mathematician, philosopher, or engineer, either in this country, or any part of the maritime world, with success. Some have taken it up in error, they not being acquainted with the art practically; others have strove to show more than the defects, being interested with chain or wire rope, &c.; few have given the art of ropemaking that place among the scientific branches it deserves, with the exception of Captain Joseph Huddart, to whose invention of the Register Plate and Tube the trade, and the whole of the maritime world, are indebted.

It must be allowed that among the many arts necessary to navigation, one on which the safety of the ship depends for the security of sails, masts and yards, and consequently the hull, and lives of the passengers and crew, is the manufacture of cordage, used as tow-lines, rigging, running gear, &c.

But it has often happened, that where no expense has been spared, the best material selected, and the greatest care taken in its manufacture, that when it is placed in the hands of the fitter, rigger, or sailor, oftentimes the properties of the rope are destroyed, and the blame put upon the manufacturer; I speak this from experience, having been employed in all the branches as ropemaker, sailor, and rigger, many years.

I at one time made the cordage for the outfit of a ship of 600 tons. I knew the materials to have been the best, and every care taken in the manufacture. The ship went her voyage, and when she returned, the captain made his report, and stated the cordage to have been the worst he had ever seen ; that a 5-inch hawser had stretched down to $2\frac{1}{2}$ inches. It being my duty to investigate into this report, I saw the captain, and he persisted in the statement. I then inquired of him how many fathoms of 5-inch rope he received on board, he answered, 130 fathoms. I then wished to know how many fathoms of $2\frac{1}{2}$ inches he had from the 130 fathoms of 5 inches—he did not know ; and when I said he should have 520 fathoms of $2\frac{1}{2}$ inches, this he could not understand, but gave me to understand that some one had offered to supply cordage cheaper, and he could not leave the firm without some excuse—there was the fault of the cordage.

This work has been written with the view of assisting the workman in obtaining a knowledge of the calculations necessary to the art of ropemaking ; having, in the course of my own practical employment, been frequently in want of such rules, and as often disappointed when asking for information of those it might have been expected from, I was, in consequence, compelled to form rules to enable me to carry on the work and to answer questions put to me by the officers of the dockyards through the Lords of the Admiralty, and which were often very absurd ; hence, the following rules and tables will be found chiefly to consist of those practical rules connected with the art of ropemaking.

TREATISE ON ROPEMAKING.

HEMP.

SEED to be sown, should be of the preceding year, because it is an oily grain, and is apt to become rancid if kept too long; it is also advisable to choose the seed every second year from a different soil.

The time for sowing is from the beginning to the end of April; if sown earlier, the plants become tender, the frost will injure, if not totally destroy them. The plants should be left thick, as without this precaution, the plants grow large, the bark woody, and the fibres harsh.

The ripeness of the male plant is known by the leaves turning yellow, and the stem of a whitish colour.

The ripeness of the female, by the opening of the pods so much, that the seed may be seen—they will have a brownish appearance.

The harvest for pulling the male is about August, the female not being fit until Michaelmas. When gathered, it is taken by the root end in large handfuls, and with a wooden sword the flowers and leaves are dressed off—twelve hands form a bundle, head, or layer. It is immersed in water as soon as possible; as by drying, the mucilage hardens, and it requires a more severe operation to develop the bark than when macerated directly, which is injurious to the fibre. If let lie in water too long, the fibres are too much divided, and by an undue dissolution of the gum, would not have the

strength to stand the effort it should, in being dressed. But
if not sufficiently steeped, it becomes harsh, coarse, non-elastic,
and encumbered with woody shives, which is a great defect.
The next operation is to separate the fibres from the stem ;
this is done by a process called scutching, formerly practised,
but now by a machine, called a brake; the operation is only
breaking the reed or woody part, for the fibre itself, of which
is the filamentous substance ; hemp only bends, and does not
break. The strength of the longitudinal fibres is superior to
the fibres by which they are joined ; or, in other words, it
requires more to break them than to separate them from one
another, as by rubbing or beating, causes the longitudinal
fibre to separate, and in proportion of a greater or less degree
of that separation, it becomes more or less fine, elastic, and
soft.

When intended for cordage or coarse yarn, it requires only
to be drawn through a coarse heckle; but if for fine yarn,
through heckles of various degrees of fineness. In this process
the pins carry off a part of the gum in the form of dust, which
is very pernicious, and by dividing the fibres, separate entirely
the heterogeneous mass. To effect this, the heckle is fixed
upon a frame, one side inclining from the workmen, who,
grasping a handful of hemp in his hands, draws it through
the heckle pins, which divides the fibres, cleanses and
straightens them, and renders the hemp fit for spinning ;
if the fibres were spun longitudinally, the yarn would be
stronger, more easily join, and require less twist.

SPINNING.

When the spinner has placed the hemp around him, he commences by taking hold of the middle of the fibres, and attaching them to the rotatory motion that supplies twist, which, upon receiving, he steps backwards, doubling the fibres in the operation. When the yarn is spun, it is warped into hauls or junks, which contain a certain number of threads or yarns in proportion to the size and weight. The hauls are then tarred, if required. The tar should be good, and of a bright colour when rubbed by the fingers—Archangel being the best; mixing with it, at times, a portion of Stockholm, to ameliorate and soften that which has been boiled, as by repeated boiling it becomes of a pitchy substance, and makes the cordage stiff, difficult to coil, and liable to break. The tar should at first be heated to a temperature of 220 degrees of Fahrenheit previous to commencing operations, so that the aqueous matter may be evaporated, and any dirt or other dense matter precipitated and taken out, thereby cleansing it from all impurities ; as the yarn, passing through the tar, takes or draws in a quantity of moisture, and the atmospheric air pressing upon the surface, has a tendency of lowering the temperature ; it never should descend while in operation below 212 degrees to evaporate that moisture. The yarn should not pass through the tar at a greater speed than fifteen feet per minute, to allow it to imbibe a sufficient quantity to prevent decay, and cause an amalgamation to take place, rendering the cordage more durable in exposed situations, weaker by its adhesion to the fibre which makes it more rigid, and destroys a small portion of its strength and elasticity. After being tarred, the hauls are left for several hours to allow any moisture to evaporate ; it is then coiled into the yarn-house, and left for several days to allow the tar to harden, and adhere more closely to the fibre; otherwise, should it be made into cordage directly after being tarred, the tar would press to the surface, decay take place in the centre, and give the cordage an unsightly appearance. When the hauls have lain a time in store, they are wound upon bobbins, the haul being stretched along the floor of a shed; and each end being formed in loops or bights, are placed upon hooks, and made taut by tackles; the workman takes the end of four yarns and separates them, passing each end through a gauge, attaches them to bobbins placed upon a machine to receive them, called a winding machine. When the bobbins are full, they each contain about 500 fathoms of yarn, or in proportion to the size of the yarn, and are taken from the machine and replaced by empty ones, and the operation proceeds.

The bobbins of yarn are then taken to a frame made to receive them, and the ends passed through a metallic plate perforated with holes in concentric circles; each yarn is passed through a single hole to the number of yarns required to form a strand; the whole are then brought together, and drawn through a cylindrical metallic tube, having a bore equal in diameter to the number of yarns when compressed. It is then attached to a machine which is drawn down the rope-walk by steam or some other power; at the same time a rotatory motion is given to twist the yarns into a strand, making an uniform cylinder. These machines are called registers, because they register the length. Forming giving a proper formation, and equalising for the equality of twist given the strands over the old method.

There are other machines for making cordage upon more scientific principles, and which give a greater uniformity of twist or angle, such as Captain Huddart's, for these reasons: —the backward travelling movement of any register, forming, or equalising machine that is or may be used in a rope-walk, the retrograde movement of such a machine towards the bottom of the walk to which the strands are drawn, and where the most improved and best principle is or may be adopted, has hitherto been found defective. The machines being worked by ropes applied in different ways, causes non-uniformity in the twist or angle ; as, in some cases, the rope is made to draw the machine by fastening one of its ends to the machine, and the other to a capstan at the bottom of the walk, the twist being given by the rotatory motion of the wheels upon which it travels ; in other cases, a rope, termed a ground-rope, made fast at each end of the walk, and having one or more turns round the barrel of the machine to give the required twist to the strands. Also an endless rope passing from one end of the walk to the other, the one end passing round a movable pulley, the other round a capstan, with power to drive the machine ; the rope is then passed round a gab-wheel upon the machine ; the capstan being put in motion, the endless rope drives the gab-wheel, and causes the machine to retrograde or travel along the ground-rope which gives motion to the pinions, and twist the strands. The great object to be obtained is in regulating the retrograding or travelling motion, and to preserve a certain speed in a given time, in order that the strands may receive a proper degree of twist in a certain length.

The next operation, the strands are made into a rope by being attached to the machines at each end of the walk, and brought to a certain degree of tension by the means of tackles ; a wood frame, called a drag, is made fast to the machine, and some heavy material placed upon it to retain that tension when released from the tackles. The machines are then put in motion, and as the strands receive tortion they shorten in their length—this is called hardening ; but from various causes, during this process, an inequality of tension takes place, one

strand becoming slack and the others tight, therefore of un-
equal lengths, although originally of equal lengths, and
received the same number of twist or turns by machines of
the most approved principle. The method practised to
remedy this, is to twist up the slack strand, making it
harder and smaller, and consequently it cannot lay evenly
in the rope, and will be the first to break. It is also
obvious that an after-twist must be given the rope to cause the
strands to unite, as for every twist given the rope the same is
taken from the strands ; hence the same number of twists the
rope receives, the same number must be given to the strands,
and any increase given the rope in making or rounding cannot
be retained, but must come out when the rope is put upon a
strain. When the strands have received a sufficient hardness
of twist, they are placed upon one hook upon one of the
machines ; a cone of wood, called a top, with grooves cut in
the surface sufficiently large to receive the strands are then
put between them ; the machines are then put in motion, the
strands made to bear equally the tails wrapped around the
rope, and all is ready for closing. The machine that twists
the rope being set so as to make two revolutions, while the
machine that twists the strands makes but one revolution ;
this extra revolution given the rope being requisite to overcome
the friction which is caused by the top, tails, and the stake
heads which are placed at every five fathoms to support the
strands and rope, and which extra revolutions cannot be retained in the rope.

RULES AND EXAMPLES.

NOTE.

SIGNS AND ABBREVIATIONS USED IN THIS WORK.

All the calculations are decimally.

$=$ Sign equal to, as 4 added 8 $=$ 12.

$+$,,   addition, as 5 plus 3 $=$ 8.

$\div$,,   division, as 8 divided 2 $=$ 4.

$\times$,,   multiplied, as 6 multiplied 3 $=$ 18.

$-$,,   subtraction, as 9 minus 6 $=$ 3.

$:::::$,,   proportion, that 2 is to 3 as 4 is to 6.

$\sqrt{}$ Square root $\left.\right\}$ the extraction of roots, thus $\sqrt{64} = 8$,
$\sqrt[3]{}$ Cube root $}$ —and $\sqrt[3]{64} = 4$.

4^2 to be squared, $\left.\right\}$ raising the powers thus, $4^2 = 16$,
4^3 to be cubed $}$ and $4^3 = 64$.

$3 + 5 \times 4 = 32$, that 3 added to 5, multiplied by 4 $=$ 32.

$\sqrt{5^2 - 3^2} = 4$ .. 5 squared less 3 squared, the square root of remainder $=$ 4.

$\dfrac{\sqrt[3]{20 \times 12}}{30} = 2$ ... 20 multiplied by 12, divided by 30 the cube $=$ 2.

$\dfrac{24 \times 6 + 12 \times 3 \times 4}{12} = 60$ ... 24 multiplied by 6, and

12 multiplied by 3, added together, multiplied by 4 and divided by 12, the quotient $=$ 60.

THREAD AND YARN.

RULE.

To find the weight of a thread of a certain length and size to make a 3-inch rope.

EXAMPLE.

One inch diameter is 3·1416 circumference, then 3·1416 ÷ 2 = 1·5708, the circumference of a strand of a 3-inch rope.

And 3·1416 : 1 :: 1·5708 : ·5 diameter
. 1·5708 ÷ 2 = ·7854 & 5 ÷ 2 = ·25, radii.
·7854 × ·25 = 19·6350, area of a strand.
19·6350 ÷ 20 = ·9817, area of a thread.

Then given the area of the strand, 20, we have, by the square root, 160 fathoms, as $\sqrt{20}$ = 4·4721, and 20 : 4·4721 :: 160 : 71·5536, and 71·5536 ÷ 20 = 3·5775 lbs. each thread. If we give the square root equal 4·5, we have a formula for all other sizes.

EXAMPLE.

20 : 4·5 :: 160 : 72 lbs., 20 threads.
72 ÷ 20 = 3·6 lbs., one thread, 160 fathoms.

Then,

72 being the formula, the same will answer for every size yarn or thread.

Weight of yarn, 160 fathoms.

White Yarn.

72 ÷ 16 = 4·5 = 16 thread yarn.
72 ÷ 18 = 4· = 18 ,,
72 ÷ 20 = 3·6 = 20 ,,
72 ÷ 25 = 2·8 = 25 ,,
72 ÷ 40 = 1·8 = 40 ,,

Tarred Yarn.

$$90 \div 16 = 5\cdot62 = 16 \text{ thread yarn.}$$
$$90 \div 18 = 5\cdot \quad = 18 \quad \text{,,}$$
$$90 \div 20 = 4\cdot5 \quad = 20 \quad \text{,,}$$
$$90 \div 25 = 3\cdot84 = 25 \quad \text{,,}$$
$$90 \div 40 = 2\cdot25 = 40 \quad \text{,,}$$

White Yarn, 200 fathoms.

$$90 \div 40 = 2\cdot25 = 40 \text{ thread yarn.}$$
$$90 \div 30 = 3\cdot \quad = 30 \quad \text{,,}$$
$$90 \div 25 = 3\cdot6 \quad = 25 \quad \text{,,}$$
$$90 \div 20 = 4\cdot5 \quad = 20 \quad \text{,,}$$
$$90 \div 18 = 5\cdot \quad = 18 \quad \text{,,}$$

Tarred Yarn, 200 fathoms.

$$112 \div 40 = 2\cdot8 \quad = 40 \text{ thread yarn.}$$
$$112 \div 30 = 3\cdot73 = 30 \quad \text{,,}$$
$$112 \div 25 = 4\cdot48 = 25 \quad \text{,,}$$
$$112 \div 20 = 5\cdot6 \quad = 20 \quad \text{,,}$$
$$112 \div 18 = 6\cdot2 \quad = 18 \quad \text{,,}$$

RULE.

To find the number of turns or twist required for any size yarn.

EXAMPLE.

The square root of the size multiplied by 375 is the number of twist per foot.

$$\sqrt{40} = 6\cdot3245 \times 3\cdot75 = 21\cdot8268 \text{ twist per foot.}$$
$$\sqrt{30} = 5\cdot4772 \times 3\cdot75 = 20\cdot5395 \quad \text{,,}$$
$$\sqrt{25} = 5\cdot \quad\quad \times 3\cdot75 = 18\cdot7500 \quad \text{,,}$$
$$\sqrt{20} = 4\cdot4721 \times 3\cdot75 = 16\cdot7703 \quad \text{,,}$$
$$\sqrt{18} = 4\cdot2426 \times 3\cdot75 = 15\cdot9097 \quad \text{,,}$$

RULE.

To find the size of the yarn by the number of threads in the haul.

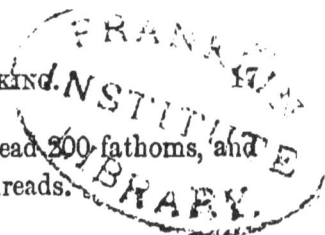

A haul of yarn of eighteens, each thread 200 fathoms, and weigh 900 lbs., white containing 180 threads.

EXAMPLE.

$90 \div 18 = 5\cdot$ & $900 \div 5\cdot = 180$ threads.
$90 \div 20 = 4\cdot5$ & $900 \div 4\cdot5 = 200$ „
$90 \div 30 = 3\cdot$ & $900 \div 3\cdot = 300$ „

RULE.

To find the number of threads in the junk; a junk of eighteens, each thread 160 fathoms, and weigh 376 lbs., and 38 blocks.

$18 = 376 \div 4\cdot = 94 \times 160 = 152 \div 4 = 38$ blocks.
$20 = 378 \div 3\cdot6 = 105 \times 160 = 168 \div 4 = 42$ „
$25 = 378 \div 2\cdot8 = 135 \times 160 = 216 \div 4 = 54$ „

Whale lines are made from dressed hemp. And one hundred and four dozen will make ten lines, each about 120 lbs. white, and 147 lbs. tarred; $10\cdot4$ dozen, one line; each thread $1\cdot55$ lbs.; 180 fathoms, ten lines; one haul of 900 threads.

N.B.—Whale lines are made of different sizes and length, as lines for the Hudson's Bay, Greenland fishery, and South Sea, therefore, the person who has the order must work accordingly.

RULE.

To find the length of yarn to make any length of three-strand shroud, or hawser laid cordage.

Multiply the length of rope by 3, and divide by 2.

EXAMPLE.

90 fathoms of rope required.

$90 \times 3 = 270 \div 2 = 135$ fathoms of yarn.
$130 \times 3 = 390 \div 2 = 195$ „

C

RULE.

Four-strand Shroud or Hawser laid.

Multiply length of rope by 11, and divide by 7.

EXAMPLE.

90 fathoms of rope required.

$90 \times 11 = 990 \div 7 = 140$ fathoms of yarn.
$130 \times 11 = 1430 \div 7 = 204$,,

RULE.

Three-strand Cable laid.

Multiply length of cable by 5, and divide by 3.

EXAMPLE.

100 fathoms of cable required.

$100 \times 5 = 500 \div 3 = 166 \cdot 6$ fathoms of yarn.
$120 \times 5 = 600 \div 3 = 200 \cdot$,,

RULE.

Four-strand Cable laid.

Multiply length of cable by 7, and divide by 4.

EXAMPLE.

30 fathoms of cable required.

$30 \times 7 = 210 \div 4 = 52$ fathoms of yarn.
$60 \times 7 = 420 \div 4 = 105$,,

Another method :—Add a cipher to the length of cable required, and divide by 6, adding to the quotient one-twelfth of the cable's length.

EXAMPLE.

60 fathoms of cable required.

60 + 0 = 600 ÷ 6 = 100 + 5 = 105 fathoms of yarn.
40 + 0 = 400 ÷ 6 = 66·6 + 5 = 71·6 ,,

RULE.

To find the length of yarn to make a heart for four-strand shroud hawser laid cordage.

Multiply the length of rope by 5, and divide by 4.

EXAMPLE.

60 fathoms of rope required.

60 × 5 = 300 ÷ 4 = 75 fathoms of yarn.
19 × 5 = 95 ÷ 4 = 23·75 ,,

RULE.

To find the length of yarn to make a heart for four-strand cable laid.

Multiply length of cable by 6, and divide by 5.

EXAMPLE.

40 fathoms of cable required.

40 × 6 = 240 ÷ 5 = 48 fathoms of yarn.
10 × 6 = 60 ÷ 5 = 12 ,,

Or add one-fifth the length required.

EXAMPLE.

40 ÷ 5 = 8 + 40 = 48 fathoms.
10 ÷ 5 = 2 + 10 = 12 ,,

c 2

RULE.

Length of yarn to make any length of bolt-rope :—
Multiply length of rope by 7, and divide by 5.

EXAMPLE.

90 fathoms of bolt-rope required.

90 × 7 = 630 ÷ 5 = 126 fathoms of yarn.
50 × 7 = 350 ÷ 5 = 70   „

Length of yarn to make any length of flat rope :—
Multiply length of flat rope by 4, and divide by 3.

EXAMPLE.

90 fathoms of flat rope.

90 × 4 = 360 ÷ 3 = 120 fathoms of yarn.
100 × 4 = 400 ÷ 3 = 133·3   „

RULE.

Length of yarn to make any length of cable-laid tacks:
Multiply length by 5, divide by 3.

EXAMPLE.

12 fathoms of tack required.

12 × 5 = 60 ÷ 3 = 20 fathoms of yarn.
 9 × 5 = 45 ÷ 3 = 15   „

TACKS.

If tapered to custom, is ⅓, or 15 + 5, equal shank 15, and
head 5 fathoms = 20 fathoms.

20 × 9 = 180 fathoms to warp the yarn for 12 fathoms of
tack, and to taper is 15 — 5 = 10 fathoms shank, and
10 × 6 = 60 ÷ 12 = 5 feet each taper.

SHEETS

Are tapered from half the length, and are made three-strand shroud laid; thus, a sheet 30 fathoms would require 45 fathoms of yarn.

EXAMPLE.

30 fathoms of sheet required.

$30 \times 3 = 90 \div 2 = 45$ fathoms of yarn.

$45 \div 2 = 22.5$ fathoms to taper.

$22.5 \times 6 = 135 \div 5 = 27$ tapers, 5 feet each.

RULE.

To find the number of yarns or threads per hook or strand for any size three-strand shroud, hawser-laid. Multiply the square of the rope by the size of the yarn, and divide by 9.

EXAMPLE.

A 6-inch rope required, and the yarn twenties.

$6^2 = 36 \times 20 = 720 \div 9 = 80$ threads per hook.

$9^2 = 81 \times 18 = 1458 \div 9 = 162$,,

RULE.

Four-strand Shroud Hawser.

Multiply the square of the rope by the size of the yarn, and divide by 27; multiply the quotient by 2.

EXAMPLE.

6-inch rope, yarn eighteens.

$6^2 = 36 \times 18 = 648 \div 27 = 24 \times 2 = 48$ thrds. per hk.

$6^2 = 36 \times 20 = 720 \div 27 = 26.26 \times 2 = 53$,,

RULE.

Three-strand Cable laid.

Multiply the square of the cable by the size of the yarn, and divide by 36.

EXAMPLE.

12-inch cable, yarn eighteens.

$12^2 = 144 \times 18 = 2592 \div 36 = 72$ threads per hook.
$16^2 = 256 \times 20 = 5120 \div 36 = 142$ „

RULE.

Four-strand Cables or Stays.

Multiply the square of the cable or stay by the size of the yarn, and divide by 48.

EXAMPLE.

15-inch cable or stay, yarn eighteens.

$15^2 = 225 \times 18 = 4050 \div 48 = 84$ threads per hook.
$9^2 = 81 \times 20 = 1620 \div 48 = 34$ „

HEARTS.

Hearts are generally made of a different sort of yarn than the rope; more soft and elastic; so that it will admit of tortion and not break.

RULE.

To find the number of threads for the heart for four-strand shroud hawser, divide the number of threads in one strand by 3, and divide the quotient by 3, will give the number of threads per hook.

Example.

12-inch shroud hawser.

Strand, $192 \div 3 = 62 \div 3 = 21$ per hook.

Rule.

Four-strand Cable or Stay.

The number of threads for the heart is the number contained in the lessom, or one-twelfth of the number contained in the cable or stay.

Example.

12-inch cable or stay.

Lessom, 58 threads ; cable or stay, 692.

$692 \div 12 = 58 \div 3 = 19$ threads per hook.

Rule.

To find the length to register the strands for any length of three-strand shroud, hawser laid.

Multiply length of rope by 7, and divide by 5.

Example.

100 fathoms of rope required.

$100 \times 7 = 700 \div 5 = 140$ fathoms to form.

$120 \times 7 = 840 \div 5 = 168$ „

Rule.

Four-strand Shroud Hawser.

Multiply length of rope by $7\frac{1}{4}$, and divide by 5.

Example.

$100 \times 7.25 = 725.00 \div 5 = 145$ fathoms to form.
$60 \times 7.25 = 435.00 \div 5 = 87$ „

Rule.

Three-strand Cable laid.

Multiply the length of cable by 3, and divide by 2.

Example.

120 fathoms of cable required.

$120 \times 3 = 360 \div 2 = 180$ fathoms to register.
$40 \times 3 = 120 \div 2 = 60$ „

Rule.

Four-strand Cable laid.

Multiply the length of cable or stay by 11, and divide by 7.

Example.

40 fathoms cable or stay required.

$40 \times 11 = 440 \div 7 = 62.8$ fathoms to register.
$12 \times 11 = 132 \div 7 = 18.8$ „

BOLT-ROPE.

Rule.

Multiply length of bolt-rope by 9, and divide by 7.

Example.

100 fathoms of bolt-rope required.

$100 \times 9 = 900 \div 7 = 128.7$ fathoms to register.
$130 \times 9 = 1170 \div 9 = 167.$ „

FLAT ROPE.

RULE.

Multiply the length of flat rope by 10, and divide by 8.

EXAMPLE.

170 fathoms of flat rope required.

170 × 10 = 1700 ÷ 8 = 212·5 fathoms to register.
90 × 10 = 900 ÷ 8 = 112·5 ,,

RULE.

Tacks Cable laid.

Multiply the length of tack by 3, and divide by 2.

EXAMPLE.

9 fathoms of tack required.

9 × 3 = 27 ÷ 2 = 13·5 fathoms to register.

To find the number of threads for the lessom, the length to register, length of head and shank, and number of tapers.

EXAMPLE.

A tack 9 fathoms 5 inches, yarn twenties.

5^2 = 25 × 20 = 500 ÷ 36 = 13·8 or 14 threads per hook.
9 × 3 = 27 ÷ 2 = 13·5 length to register.
13·5 × 2 = 27 ÷ 3 = 9 ,, of shank.
13·5 − 9 = 4·5 ,, of head.
14 × 3 = 42 ÷ 5 = 8 number of tapers.
9 × 6 = 54 ÷ 8 = 6·75 feet each taper.

RULE

To find what should be the size of the strand to make a three-strand shroud hawser.

Multiply size of rope by 3, and divide by 6.

EXAMPLE.

8-inch rope required.

$8 \times 3 = 24 \div 6 = 4$ inches, size of strand.
$6 \times 3 = 18 \div 6 = 3$ „   „

RULE.

Four-strand Shroud Hawser.

Multiply size of shroud hawser by 6, and divide by 15.

EXAMPLE.

9-inch shroud hawser required.

$, 9 \times 6 = 54 \div 15 = 3\cdot6$ inches, size of strand.
$5 \times 6 = 30 \div 15 = 2\cdot$ „   „

RULE.

Three-strand Cable.

Multiply size of lessom by 6, and divide by 24.

EXAMPLE.

24-inch cable required.

$24 \times 6 = 144 \div 24 = 6$ inches, size of lessom.

RULE.

Three-strand Cable.

Size of strand.

Multiply size of cable by 6, and divide by 12.

EXAMPLE.

24-inch cable strand required.

$24 \times 6 = 144 \div 12 = 12$ inches, size of strand.
$7 \times 6 = \ 42 \div 12 = \ 3\cdot5$ „   „

Rule.

Four-strand Cable or Stay.

Size of lessom.

Multiply size of cable by 2, and divide by 9.

Example.

12-inch cable or stay.

$12 \times 2 = 24 \div 9 = 2\cdot66$ inches, size of lessom.

$14 \times 2 = 28 \div 9 = 3\cdot1$ „ „

Rule.

Four-strand Cable or Stay.

Size of strand.

Multiply size of cable by 4, and divide by 9.

Example.

9-inch cable or stay.

$9 \times 4 = 36 \div 9 = 4\cdot$ inches, size of strand.

$15 \times 4 = 60 \div 9 = 6\cdot6$ „ „

Rule.

To find what length of rope will make a strand for a length of three-strand cable.

Multiply length of cable by 10, and divide by 9.

Example.

90 fathoms of cable.

$90 \times 10 = 900 \div 9 = 100$ fathoms of rope.

$120 \times 10 = 1200 \div 9 = 133$ „

Rule.

Four-strand Cable or Stay.

Multiply length of cable or stay by 5, and divide by 4.

Example.

40 fathoms of cable or stay.

$40 \times 5 = 200 \div 4 = 50$ fathoms of rope.
$8 \times 5 = 40 \div 4 = 10$ „   „

Rule.

To find the length of a rope when stretched by the circumference.

Multiply the square of the rope before the strain, and divide by the square of the rope when stretched.

Example.

A 6-inch rope, 60 fathoms, stretched to 3 inches.

$6^2 = 36 \times 60 = 2160$ and $3^2 = 9$.
$2160 \div 9 = 240 \div 4 = 4$ times the length.

Rule.

Three-strand Shroud Hawser.

To find the diameter of tube for any size.
Multiply size of rope by 5, and divide by 3.

Example.

A 9-inch rope required.

$9 \times 5 = 4{\cdot}5 \div 3 = 1{\cdot}5$-inch diameter.
$4 \times 5 = 2{\cdot}0 \div 3 = {\cdot}66$ „

RULE.

Four-strand Shroud Hawser.

Multiply size of rope by 10, and divide by 7.

EXAMPLE.

An 8-inch rope required.

$8 \times 10 = 8·0 \div 7 = 1·14$-inch diameter.

$5 \times 10 = 5·0 \div 7 = ·71$ „

RULE.

Three-strand Cable.

Multiply size of cable by 10, and divide by 12.

EXAMPLE.

A 20-inch cable required.

$20 \times 10 = 20·0 \div 12 = 1·66$-inch diameter.

$12 \times 10 = 12·0 \div 12 = 1·$ „

RULE.

Four-strand Cable or Stay.

Multiply size of cable or stay by 5, and divide by 7.

EXAMPLE.

A 14-inch cable or stay.

$14 \times 5 = 7·0 \div 7 = 1·0$-inch diameter.

$9 \times 5 = 4·5 \div 7 = ·64$ „

RULE.

Three-strand Cable.

To find the weight of 120 fathoms.

Square the size of the cable, and divide by 4.

EXAMPLE.

A 12-inch cable.

$12^2 = 144 \div 4 = 36$ cwt.—120 fathoms.

$6^2 = 36 \div 4 = 9$ „   „

Weight of 1 fathom,

Multiply the square of the cable by 15, divide by 64.

EXAMPLE.

A 12-inch cable.

$12^2 = 144 \times 15 = 2160 \div 64 = 33\cdot6$ lb.

$20^2 = 400 \times 15 = 6000 \div 64 = 93\cdot75$ lb.

SHROUD HAWSERS.

The weight of these ropes is to each other as the square of their circumferences, or as 3 to 324 pounds; so is any other size.

EXAMPLE.

6-inch shroud hawser.

$3^2 = 9 : 324 :: 9^2 = 36 : 1296$ lb. — 130 fathoms.

$3^2 = 9 : 324 :: 9^2 = 81 : 2916$ lb.   „

STRENGTH OF CORDAGE.

The strength of ropes of the same lay is in proportion to the number of yarns or threads, but three-strand will support more by about one-sixth than when laid into a cable.

EXAMPLE.

An 8½ three-strand cable.

$8\cdot5^2 = 7225 \times 20 = 1445 \div 36 = 40 \times 9 = 360$ threads.

A 7½ shroud hawser.

$$7 \cdot 5^2 = 525625 \times 20 = 105125 \div 9 = 117 \times 3 = 351 \text{ thrds.}$$

Showing nine threads less in the shroud hawser than in the cable, and the shroud hawser 1 inch less in size; and as twist diminishes strength, the hawser will be stronger.

Cable 200 fathoms of yarn, 120 fathoms cable.
Hawser 195 „ 130 „

RULE.

To calculate the strength of three-strand hawsers. Square the size of rope, and divide by 3.

EXAMPLE.

8 and 9-inch ropes.

$$9^2 = 81 \div 3 = 27 \text{ tons to break it.}$$
$$6^2 = 36 \div 3 = 12 \quad \text{„}$$

Three-strand Cable.

Square the size of the cable, and divide by 5.

EXAMPLE.

22½ and 17½-inch cables.

$$22 \cdot 5^2 = 506 \cdot 25 \div 5 = 101 \cdot 25 \text{ tons to break.}$$
$$17 \cdot 5^2 = 306 \cdot 25 \div 5 = 61 \cdot 25 \quad \text{„}$$

Hitherto there has been no practical rule to calculate the strength of cordage, therefore I have given those formulas I have found to be correct, from practice, an actual test.

Four-strand cordage is of considerably less strength than three-strand, upon account of the additional hard or twist it receives in the making, four-strand shroud hawser being stronger than four-strand cable; one-thirteenth of the number of the yarns that compose a four-strand rope is a base called a heart, and forms a centre to the rope; the strands being wound spirally round the heart, it becomes a radius, and when the rope is put upon a breaking strain the heart will be the first to break; and where the fracture takes place, the strands lose their support and the rope must break.

As the cohesive strength of a body is that force with which it resists separation in the direction of its length, no reason can be assigned why the strength should not vary directly as the section of fracture, except so far as the weight of the body may increase the force applied. But supposing the body uniform in all its parts, the strength of ropes exposed to strain longitudinally is in proportion to their transverse area.

It often happens, and most unaccountably, that ships have parted their cables and shrouds without any visible or apparent cause, even where there has been no expense spared to procure the best materials, through want of that scientific knowledge necessary to draw a line of demarcation between the maximum and minimum of strength and durability requisite in the manufacture of cordage.

WIRE-ROPE.

Report on the test of Wire-rope, 16th October, 1848.

A 3-inch wire-rope spliced to a 7½ hemp-rope shroud, shroud broke at 11¾ tons; wire-rope uninjured.

No. 2.

A 3½-inch wire-rope, spliced to an 8-inch hemp-rope, shroud broke at 10½ tons ; wire-rope uninjured.

This report must have been made for interest, and detrimental to the hemp-ropemakers generally.

EXAMPLE.

Diameter of 7·5 = $\sqrt{2·4}$ = 57· diameter 8 = $\sqrt{2·5}$ = 65.
And 11·75 : 57 :: 65 : 13·4 tons the 8-inch.

Whereas the 8-inch should have reached 22 tons.

Practical test of wire-rope.

2⅞	16 gauge	126	wire	salvage	..	12·2·2·6
2⅞	16 „	126	„	rope	..	10·14·2·12
3¾₁₆	16 „	168	„	salvage	..	15·12·1·19
3	16 „	112	„	4-strand rope		9·3·3·19
3¼	16 „	159	„	rope	..	13·10·0·0
3⅛	16 „	150	„	„	..	11·0·1·4
2⅝	16 „	105	„	„	..	7·18·1·0

PRESS.

The press or weight placed upon the drags after the strands are properly stretched, depends, in a great measure, upon the state of the ground and the atmosphere ; the rule generally is, one barrel of 3 cwt. to every 6 threads the strand contains ; the yarn twenties, bolt-rope, strands and some other, is left to the layer.

RULE.

To find the circumference of strand or lessom length of yarn in one turn, and number of threads or yarns in each circle.

D

EXAMPLE.

Given A. 35° angle.
B. 50·yarns in the outer circle.

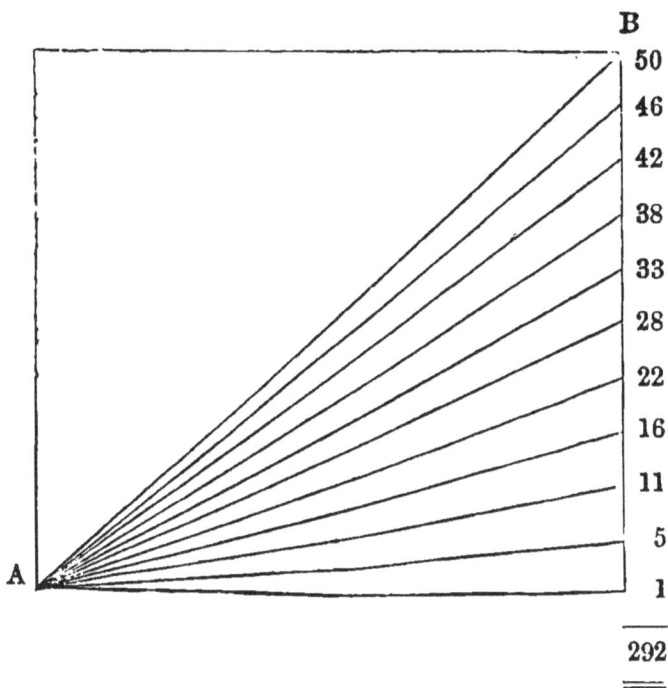

	B
	50
	46
	42
	38
	33
	28
	22
	16
	11
	5
A	1
	292

As Cosine > 35°	..	..	..	..	..	..	9·91336
: Radius 90	..	..	..	..	..	..	10
:: B Yarns 50	..	..	..	..	..	..	1·69897
: 6·104 Circumference	..	..	..	..	..	..	1·78561

As Tangent A > 35°	..	..	..	..	..	9·84522
: Radius 90	..	..	..	..	..	10·
:: D = Circumference 6·104	..	..	..	..	1·78560
: E = Length of turn 8·717	..	..	..	..	1·94038

As Radius .. 90 .. .. 10·
: B = 50 yarns 50 .. .. ·69897
:: Cosine > .. 35° .. .. 9·91336

: D = Circumference 6·104 .. .. .. 1·78561

As Radius .. 90 .. .. 10·
: D = Circumference 6·104 .. .. .. 1·78560
:: Tangent > 35° .. .. 9·84522

: E = Length of turn 8·717 .. .. .. 1·94038

Therefore,

As D = Circumference 6·104 .. .. .. 1·78560
: Radius .. 90 .. .. 10·
:: E = Length of turn 8·717 .. .. .. 1·94037

: Tangent > 35° .. .. 9·84523

As Cosine > 35° .. .. 9·91336
: D = Circumference 6·104 .. .. .. 1·78560

:: E = 50 yarns 1st circle .. .. ·69896

As D = C¹ .. 54·7 .. .. 1·73798
: Radius .. 90 .. .. 10·
:: E = length of turn 8·717 .. .. .. 1·94037

: Tangent > 32° 6′ .. .. 9·79761

As Cosine > .. 32° 6′ .. .. 9·92794
: D = C¹ .. 54·7 .. .. 1·73798

:: E = 46·33 = 46 yarns 2nd circle .. .. ·66592

As D = C² .. 48·4 .. .. 1·68484
: Radius .. 90 .. .. 10·
:: E = length of turn 8·717 .. .. .. 1·94037

: Tangent > 29° 2′.. .. .. 9·74447
_____

As Cosine > .. 29° 2′ .. .. 9·94167
: D = C² .. 48·4 .. .. 1·68484

:: E = 42·32 = 42 yarns 3rd circle .. .. ·62651
_____

As D = C³ .. 42·1 .. .. 1·62428
: Radius .. 90 .. .. 10·
:: E = length of turn 8·717 .. .. .. 1·94037

: Tangent > 25° 46′ .. .. 9·68391
_____

As Cosine > .. 25° 46′ .. .. 9·95451
: D = C³ .. 42·1 .. .. 1·62428

:: E = 37·94 = 38 yarns 4th circle .. .. ·57879
_____

As D = C⁴ .. 35·8 .. .. 1·55387
: Radius .. 90 .. .. 10·
:: E = length of turn 8·717 .. .. .. 1·94037

: Tangent > 22° 19′ .. .. 9·61350
_____

As Cosine > .. 22° 19′ .. .. 9·96618
: D = C⁴ .. 35·8 .. .. 1·55389

:: E = 33·12 = 33 yarns 5th circle .. .. ·52007
_____

As D = C⁵ .. 29·5 .. .. 1·46982
: Radius .. 90 .. .. 10·
:: E = length of turn 8·717 .. .. .. 1·94037

: Tangent > 18° 42′ .. .. 9·52945
_____

As Cosine $>$ .. 18° 42′ .. .. 9·97644
: D = C⁵ .. 29·5 .. .. 1·46982

:: E = 27·94 = 28 yarns 6th circle .. .. ·44626

As D = C⁶ .. 23·2 .. .. 1·36548
: Radius .. 90 .. .. 10·
:: E = length of turn 8·717 .. .. .. 1·94037

: Tangent $>$ 14° 43′ .. .. 9·41911

As Cosine $>$ .. 14° 43′ .. .. 9·98552
: D = C⁶ .. 23·2 .. .. 1·36548

:: E = 22·44 = 22 yarns 7th circle .. .. ·35100

As D = C⁷ .. 16·9 .. .. 1·22787
: Radius .. 90 .. .. 10·
:: E = one turn 8·717 .. .. 1·94037

: Tangent $>$ 10° 58′ .. .. 9·28750

As Cosine $>$ .. 10° 58′ .. .. 9·99199
: D = C⁷ .. 16·9 .. .. 1·22789

:: E = 16·6 = 16 yarns 8th circle .. .. ·21988

As D = C⁸ .. 10·6 .. .. 1·02530
: Radius .. 90 .. .. 10·
:: E = one turn 8·717 .. .. 1·94037

: Tangent $>$ 6° 56′ .. .. .. · .. 9·08493

As Cosine $>$ ..   6° 56'   .. ..   9·99681
: D = C⁸   .. 10·6 .. ..   1·02530

:: E = 10·5 = 11 yarns 9th circle .. ..   ·02011

Then $D = C^{94}$: 745, circumference of the 10th circle, the remaining vacancy.

Therefore,

745² × ·07958, the area of unity, gives 5·653, and completes a strand of 292 threads or yarns.

RULE.

To find the length of yarn in each series of circles or shells.

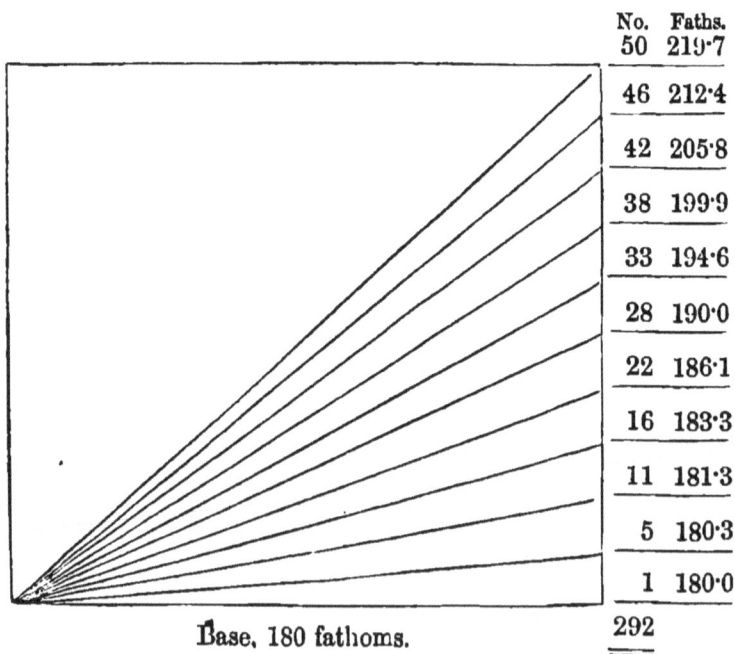

	No.	Faths.
	50	219·7
	46	212·4
	42	205·8
	38	199·9
	33	194·6
	28	190·0
	22	186·1
	16	183·3
	11	181·3
	5	180·3
	1	180·0
Base, 180 fathoms.	292	

As Radius	..	90 ..	..	..	..	..	10·
′ : Secant	..	35°	..	..	..	..	10·08664
:: Base ..	..	180 ..	..	..	..	..	2·25527
: Length of yarn 219·7 ,			..	..	..	..	2·34191

As Radius	..	90 ..	..	..	..	..	10·
: Secant	..	32° 6′	..	..	..	..	10·07205
:: Base ..	..	180 ..	..	..	..	..	2·25527
: Length of yarn 212·4			..	..	..	..	2·32732

As Radius	..	90 ..	..	..	..	..	10·
: Secant >	..	29° 2′	..	..	..	..	10·05832
:: Base ..	..	180 ..	..	..	..	..	2·25527
: Length of yarn 205·8			..	..	..	..	2·31359

As Radius	..	90 ..	..	..	..	..	10·
: Secant >	..	25° 46′	..	..	..	..	10·04548
:: Base ..	..	180 ..	..	..	..	..	2·25527
: Length of yarn 199·9			..	..	..	..	2·30075

As Radius	..	90 ..	..	..	..	..	10·
: Secant >	..	22° 19′	..	..	..	..	10·03381
:: Base ..	..	180 ..	..	..	..	..	2·25527
: Length of yarn 194·6			..	..	..	..	2·28908

As Radius	..	90 ..	..	..	..	..	10·
: Secant >	..	18° 42′	..	..	..	..	10·02355
:: Base ..	..	180 ..	..	..	..	..	2·25527
: Length of yarn 190 ..			..	..	..	..	2·27882

As Radius	..	90	..	..	..	..	..	10·
: Secant >	..	14° 43′	..	..	..	..		10·01449
:: Base	.. ..	180	..	..	..	..	..	2·25527
: Length of yarn 186·1			..	..	..	..		2·26976

As Radius	..	90	..	..	..	..	..	10·
: Secant >	..	10° 58′	..	..	..	..		10·00800
:: Base	.. ..	180	..	..	..	..	..	2·25527
: Length of yarn 183·3			..	..	..	..		2·26327

As Radius..	..	90	..	..	..	..	..	10·
: Secant >	..	6° 56′	..	..	..	..		10·00319
:: Base	.. ..	180	..	..	..	..	..	2·25527
: Length of yarn 181·3			..	..	..	..		2·25846

As Radius	..	90	..	..	..	..	..	10·
: Secant >	..	3° 5′	..	..	..	..		10·00063
:: Base	.. ..	180	..	..	..	..	..	2·25527
: Length of yarn 180·3			..	..	..	..		2·25590

As radius : radius :: base : base, 180; and completes the strand.

RULE.

To find the angle of the strands or readies, after being hardened or shortened 12 fathoms, being registered at an angle of 27 degrees, and 180 fathoms.

As length of strand 180 .. ..   2·25527
: Radius .. .. 90 .. ..   10·
:: Sine > .. 27° .. ..   9·65705

: 39·16 .. ..   2·59822

Then,

$$39·16 - 12 = 27·16.$$

Therefore,

As length at hard 168 .. ..   2·22531
: Radius .. .. 90 .. ..   10·
:: 27·16 .. .. 27·16 .. ..   2·43393

: Sine angle .. 38° 12′ .. .. ..   9·79138

RULE.

To find the angle the rope is laid with. As length of strand hard to radius so is length of rope to the cosine angle it is laid with.

EXAMPLE.

As length of strand 165 .. ..   2·217484
: Radius .. .. 90 .. ..   10·
:: Length of rope 130 .. ..   2·113943

: Cosine angle 38° .. ..   9·896459

And,

As sine angle .. 52° .. ..   9·896530
: Radius .. .. 90 .. ..   10·
:: Length of rope 130° .. ..   2·113943

: Length of strand 165 .. ..   2·217413

Cables and Hawsers.

To find the hard length of strand divided by 15 equal hard.

EXAMPLE.

Strand 180 fathoms,

$$180 \div 15 = 12 = 12 \text{ fathoms hard.}$$

Bolt-rope.

Length of strand divided by 19 equal hard.

EXAMPLE.

$$180 \div 19 = 9\text{·}5 = 9\tfrac{1}{2} \text{ fathoms hard.}$$

The hardness of the rope depends in a great measure how the order is given. Some require the rope harder than others, therefore it is with the foreman or layer to exercise his judgment.

THREE-STRAND SHROUD HAWSERS.

TABLE showing size of Rope, size of Yarn, and the number of Threads or Yarns in a Strand and Rope of Three-strand Shroud Hawsers.

Size of Rope.	Eighteens.		Twenties.		Twenty-fives.	
	Threads in Strand.	Threads in Rope.	Threads in Strand.	Threads in Rope.	Threads in Strand.	Threads in Rope.
Inches.						
1	2	6	3	9	4	12
1¼	4	12	5	15	6	18
1½	6	18	7	21	7	21
1¾	7	21	8	24	9	27
2	8	24	9	27	11	33
2¼	10	30	11	33	14	42
2½	12	36	14	42	17	51
2¾	15	45	17	51	21	63
3	18	54	20	60	25	75
3¼	21	63	24	72	29	87
3½	25	75	27	81	34	102
3¾	28	84	31	93	39	117
4	32	96	35	105	44	132
4¼	36	108	40	120	50	150
4½	40	120	45	135	56	168
4¾	45	135	50	150	62	186
5	50	150	56	168	70	210
5¼	55	165	62	186	77	231
5½	60	180	67	211	84	252
5¾	66	198	73	219	92	276
6	72	216	80	240	100	300
6¼	78	234	87	261	108	324
6½	84	252	94	282	116	348
6¾	91	273	101	303	128	384
7	98	294	108	324	136	408
7¼	105	315	116	348	146	438
7½	112	336	125	375	156	468
7¾	120	360	134	402	167	501

THREE-STRAND SHROUD HAWSERS—*continued*.

Size of Rope.	Eighteens.		Twenties.		Twenty-fives.	
	Threads in Strand.	Threads in Rope.	Threads in Strand.	Threads in Rope.	Threads in Strand.	Threads in Rope.
Inches.						
8	128	384	142	426	178	534
8¼	136	408	151	453	189	567
8½	144	432	160	480	200	600
8¾	153	459	170	510	212	636
9	162	486	180	540	225	675
9½	180	540	200	600	250	750
10	200	600	222	666	278	834
10½	220	660	248	744	309	927
11	240	720	268	804	336	1,008
11½	264	792	292	876	367	1,101
12	288	864	320	960	400	1,200

FOUR-STRAND SHROUD HAWSER.

TABLE showing the size of Rope, size of Yarn, and number of Threads or Yarns in a Strand and Rope of Four-strand Shroud Hawser.

Size of Rope.	Eighteens.		Twenties.		Twenty-fives.	
	Threads in Strand.	Threads in Rope.	Threads in Strand.	Threads in Rope.	Threads in Strand.	Threads in Rope.
Inches.						
2	5	20	6	24	7	28
2¼	7	28	8	32	9	36
2½	8	32	9	36	11	44
2¾	10	40	10	44	14	56

Four-strand Shroud Hawsers—*continued.*

Size of Rope.	Eighteens.		Twenties.		Twenty-fives.	
	Threads in Strand.	Threads in Rope.	Threads in Strand.	Threads in Rope.	Threads in Strand.	Threads in Rope.
Inches.						
3	12	48	13	52	17	68
3¼	14	56	15	60	20	80
3½	16	64	18	72	23	92
3¾	18	72	21	84	26	104
4	21	84	23	92	30	120
4¼	24	96	26	104	34	136
4½	27	108	30	120	38	152
4¾	30	120	33	132	42	168
5	33	132	37	148	47	188
5¼	36	144	40	160	51	204
5½	40	160	44	176	56	224
5¾	44	176	48	192	61	244
6	48	192	53	212	67	268
6¼	52	208	57	228	72	288
6½	56	224	62	248	78	312
6¾	60	240	67	268	84	336
7	65	260	72	288	91	364
7¼	70	280	77	308	98	392
7½	75	300	83	332	105	420
7¾	80	320	88	352	112	448
8	85	340	94	376	119	476
8½	96	384	106	424	134	536
9	108	432	119	476	150	600
9½	120	480	132	528	168	672
10	133	532	148	592	184	736
10½	147	588	162	648	205	820
11	161	644	178	712	224	896
11½	178	712	195	780	245	980
12	192	768	213	852	267	1,068

THREE-STRAND CABLE.

TABLE showing the size of Cable, size of Yarn, and number of Threads or Yarns in the Lessom, Strand, and Cable.

Size of Cable.	Eighteens.			Twenties.		
	Threads in Lessom.	Threads in Strand.	Threads in Cable.	Threads in Lessom.	Threads in Strand.	Threads in Cable.
Inches.						
3	5	15	45	6	18	54
3½	6	18	54	7	21	63
4	8	24	72	9	27	81
4½	10	30	90	11	33	99
5	13	39	117	14	42	126
5½	15	45	135	17	51	153
6	18	54	162	20	60	180
6½	21	63	189	23	69	207
7	24	72	216	27	81	249
7½	28	84	252	31	93	279
8	32	96	288	35	105	315
8½	36	108	324	40	120	360
9	40	120	360	45	135	405
9½	45	135	405	50	150	450
10	50	150	450	55	165	495
10½	55	165	495	61	183	549
11	60	180	540	67	201	603
11½	66	198	594	73	219	657
12	72	216	648	80	240	720
12½	78	234	702	87	261	783
13	84	252	756	94	282	846
13½	91	273	819	102	306	918
14	98	294	882	110	330	990
14½	105	315	945	118	354	1,062
15	112	336	1,008	126	378	1,134
15½	120	360	1,080	134	402	1,206
16	128	384	1,152	144	432	1,296
16½	136	408	1,224	153	459	1,377
17	144	432	1,296	162	486	1,458

THREE-STRAND CABLE—*continued.*

Size of Cable.	Eighteens.			Twenties.		
	Threads in Lessom.	Threads in Strand.	Threads in Cable.	Threads in Lessom.	Threads in Strand.	Threads in Cable.
Inches.						
17½	153	459	1,377	172	516	1,548
18	162	486	1,458	182	546	1,638
18½	171	513	1,539	192	576	1,728
19	180	540	1,620	202	606	1,818
19½	190	570	1,710	212	636	1,908
20	200	600	1,800	222	666	1,998

FOUR-STRAND CABLE OR STAY.

TABLE showing size of Cable, size of Yarn, number of Threads or Yarns in Lessom, Strand, and Cable or Stay.

Size of Cable.	Eighteens.			Twenties.		
	Threads in Lessom.	Threads in Strand.	Threads in Cable.	Threads in Lessom.	Threads in Strand.	Threads in Cable.
Inches.						
3	3	9	36	4	12	48
3½	4	12	48	5	15	60
4	6	18	72	7	21	84
4½	7	21	84	8	24	96
5	9	27	108	10	30	120
5½	11	33	132	13	39	156
6	13	39	156	15	45	180
6½	15	45	180	17	51	204

FOUR-STRAND CABLE OR STAY—*continued.*

Size of Cable.	Eighteens.			Twenties.		
	Threads in Lessom.	Threads in Strand.	Threads in Cable.	Threads in Lessom.	Threads in Strand.	Threads in Cable.
Inches.						
7	18	54	216	20	60	240
7½	21	63	252	23	69	276
8	24	72	288	27	81	324
8½	27	81	324	30	90	360
9	30	90	360	34	102	408
9½	33	99	396	37	111	444
10	37	111	444	41	123	492
10½	40	120	480	45	135	540
11	45	135	540	49	147	588
11½	49	147	588	54	162	648
12	54	162	648	60	180	720
12½	58	174	696	65	195	780
13	62	186	744	70	210	840
13½	67	201	804	76	228	912
14	72	216	864	82	246	984
14½	77	231	924	88	264	1,056
15	83	249	996	94	282	1,128
15½	89	267	1,068	101	308	1,212
16	95	285	1,140	108	324	1,296

WHITE AND TARRED SHROUD HAWSERS.

TABLE showing the weight of one fathom and of one hundred and thirty fathoms of White and Tarred Shroud Hawsers.

Size.	White.					Tarred.				
	Weight of one Fathom.	Weight of 130 Fathoms.				Weight of one Fathom.	Weight of 130 Fathoms.			
Inches.		Ton.	cwt.	qrs.	lbs.		Ton.	cwt.	qrs.	lbs.
1	·22	0	0	1	1	·28	0	0	1	8
1½	·51	0	0	2	10	·63	0	0	2	26
2	·88	0	1	0	3	1·1	0	1	1	4
2½	1·38	0	1	2	12	1·7	0	2	0	1
3	2·00	0	2	1	6	2·5	0	2	3	16
3½	2·72	0	3	0	17	3·4	0	3	2	22
4	3·54	0	4	0	13	4·4	0	5	0	16
4½	4·4	0	5	0	17	5·5	0	6	1	21
5	5·5	0	6	1	20	6·9	0	8	0	4
5½	6·7	0	7	3	3	8·4	0	9	2	25
6	7·9	0	9	1	1	9·9	0	11	2	8
6½	9·3	0	10	3	13	11·7	0	13	2	9
7	10·8	0	12	2	10	13·5	0	15	2	26
7½	12·5	0	14	1	24	15·6	0	18	0	9
8	14·2	0	16	1	23	17·7	1	0	2	8
8½	16·0	0	18	2	9	20·0	1	3	0	25
9	17·9	1	0	3	9	22·4	1	6	0	4
9½	20·0	1	3	0	23	25·0	1	9	0	1
10	22·1	1	5	2	24	27·7	1	12	0	16
10½	24·4	1	8	1	11	30·5	1	15	1	21
11	26·8	1	11	0	13	33·5	1	18	3	16
11½	29·3	1	14	0	2	36·6	2	2	2	2
12	31·9	1	17	0	3	39·9	2	6	1	4

E

WHITE AND TARRED CABLES.

TABLE showing the weight of one fathom and of one hundred and twenty fathoms of White and Tarred Cables.

Size of Cable.	White.				Tarred.				
	Weight of one Fathom.	Weight of 120 Fathoms.			Weight of one Fathom.	Weight of 120 Fathoms.			
Inches.		Ton.	cwt.	qrs.	lbs.		Ton.	cwt.	qrs.	lbs.
3	1·7	0	1	3	6	2·0	0	2	1	0
3½	2·3	0	2	1	21	2·9	0	3	0	7
4	2·9	0	3	0	22	3·7	0	4	0	0
4½	3·8	0	4	0	11	4·7	0	5	0	7
5	4·6	0	5	0	0	5·8	0	6	1	0
5½	5·6	0	6	0	6	7·0	0	7	2	7
6	6·7	0	7	0	22	8·4	0	9	0	0
6½	7·8	0	8	1	14	9·9	0	10	2	7
7	8·5	0	9	0	10	11·4	0	12	1	0
7½	10·5	0	11	1	0	13·1	0	14	0	7
8	11·9	0	12	3	6	14·9	0	16	0	0
8½	12·7	0	14	1	22	16·8	0	18	0	7
9	15·0	0	16	0	22	18·7	1	1	0	0
9½	17·0	0	18	0	6	21·0	1	2	2	7
10	18·7	1	0	0	0	23·3	1	5	0	0
10½	20·6	1	2	0	6	25·7	1	7	2	7
11	22·6	1	4	0	22	28·2	1	10	1	0
11½	24·7	1	6	1	22	30·9	1	13	0	7
12	26·9	1	8	3	6	33·6	1	16	0	0
12½	29·0	1	11	1	0	36·4	1	19	0	7
13	31·5	1	13	3	6	38·5	2	2	1	0
13½	34·0	1	16	1	22	42·5	2	5	2	7
14	36·4	1	19	0	22	45·7	2	9	0	0
14½	39·3	2	2	0	6	49·0	2	12	2	7
15	42·0	2	5	0	0	52·5	2	16	1	0
15½	44·8	2	8	0	6	56·0	3	0	0	7

WHITE AND TABRED CABLES—*continued.*

Size of Cable.	White.			Tarred.		
	Weight of one Fathom.	Weight of 120 Fathoms.		Weight of one Fathom.	Weight of 120 Fathoms.	
Inches.		Ton. cwt. qrs. lbs.			Ton. cwt. qrs. lbs.	
16	47·9	2 11 0 22		59·1	3 4 0 0	
16½	50·8	2 14 1 22		63·5	3 8 0 7	
17	54·0	2 17 3 6		67·4	3 12 1 0	
17½	57·2	3 1 1 0		71·4	3 16 2 7	
18	60·5	3 4 3 6		75·6	4 1 0 0	
18½	64·0	3 8 1 22		79·8	4 5 0 7	
19	67·4	3 12 0 23		84·2	4 10 1 0	
19½	71·0	3 16 0 6		88·7	4 15 0 7	
20	74·8	4 0 0 0		93·3	5 0 0 0	

SHROUD HAWSERS.

TABLE showing size of Rope, number of Men to make it, and number of Threads each Man, as pay.

Size.	Number of Men.	Threads each Man.	Number of Threads.		Size.	Number of Man.	Threads each Men.	Number of Threads.	
Inches.			qrs. ths.		Inches.			qrs. ths.	
6ths	5½	2	1 5		2¾	11	3	5 3	
8ths	6½	2	2 1		3	12	4	8 0	
1	6½	2	2 1		3¼	12	4	8 0	
1¼	6½	2	2 1		3½	10	6	10 0	
1½	7	2	2 2		3¾	11	6	11 0	
1¾	8	2	2 4		4	12	6	12 0	
2	9	2	3 0		4¼	12	6	12 0	
2¼	9	3	4 3		4½	14	6	14 0	
2½	10	3	5 0		4¾	14	6	14 0	

E 2

Shroud Hawsers—*continued*.

Size.	Number of Men.	Threads each Man.	Number of Threads.		Size.	Number of Men.	Threads each Man.	Number of Threads.	
Inches.			qrs.	ths.	Inches.			qrs.	ths.
5	16	8	21	2	8½	30	12	60	0
5¼	16	8	21	2	9	33	15	82	3
5½	18	8	24	0	9½	36	15	90	0
5¾	18	8	24	0	10	39	15	97	3
6	22	12	44	0	10½	41	15	102	3
6½	24	12	48	0	11	43	15	107	3
7	25	12	50	0	11½	46	15	115	0
7½	28	12	56	0	12	50	18	150	0
8	30	12	60	0					

N.B.—The Layer is included.

CABLES.

Table showing size of Cable, number of men to make Strands and Cable, number of Threads each Man, as pay.

Size.	Men on Strand.	Threads each Man.	Men on Cable.	Threads each Man.	Total Number of Threads.		Size.	Men on Strand.	Threads each Man.	Men on Cable.	Threads each Man.	Total Number of Threads.	
In.					qrs.	ths	In.					qrs.	ths.
1	5	2	5	2	3	2	6	10	4	14	6	20	0
1½	5	2	5	2	3	2	6½	11	4	15	6	22	2
2	5	2	5	2	3	2	7	11	4	17	6	24	2
2½	5	2	5	2	3	2	7½	9	6	18	6	27	0
3	6	2	8	3	6	0	8	10	6	22	6	32	0
3½	6	2	8	6	10	0	8½	11	6	24	6	35	0
4	7	2	10	6	12	2	9	11	6	27	8	47	2
4½	8	2	11	6	13	4	9½	12	6	27	8	48	0
5	9	2	12	6	15	0	10	13	6	30	8	53	0
5½	9	3	13	6	17	4	10½	14	6	30	8	54	0

CABLES—continued.

Size.	Men on Strand.	Threads each Man.	Men on Cable.	Threads each Man.	Total Number of Threads.	Size.	Men on Strand.	Threads each Man.	Men on Cable.	Threads each man.	Total Number of Threads.
In.					qrs. ths.	In.					qrs. ths.
11	15	6	33	8	59 0	16	28	8	55	12	147 2
11½	16	6	33	8	60 0	16½	30	8	60	12	160 0
12	18	6	37	8	67 2	17	30	8	62	12	164 0
12½	19	6	37	8	68 2	17½	31	8	62	12	165 2
13	21	8	42	12	112 0	18	31	8	66	12	173 2
13½	22	8	42	12	113 2	18½	35	8	66	12	178 4
14	23	8	46	12	126 2	19	35	10	71	18	271 2
14½	25	8	46	12	127 4	19½	39	10	71	18	278 0
15	25	8	51	12	135 2	20	39	10	77	18	296 0
15½	26	8	53	12	140 4						

Layer included.

STRANDS AND STAY.

Table showing size of Stay, number of Men to make Strands and Stay, and number of Threads each Man, as pay.

Size.	Strands.		Total Number.	Stay.		Total Number.
	Number of Men.	Threads each Man.		Number of Men.	Threads each Man.	
Inches			qrs. ths.			qrs. ths.
4	9	2	3 0	26	4	17 2
4½	10	2	3 2	27	4	18 0
5	11	3	5 3	27	4	18 0
5½	12	3	6 0	28	4	18 4
6	13	3	6 3	28	4	18 4
6½	14	3	7 0	29	6	29 0
7	16	3	8 0	29	6	29 0
7½	18	3	9 0	30	6	30 0
8	20	3	10 0	30	6	30 0
8½	21	3	10 3	32	6	32 0
9	22	4	14 4	32	6	32 0
9½	23	4	15 2	34	6	34 0
10	25	4	16 4	34	6	34 0

TACKS AND SHEETS.

TABLE showing size of Tacks and Sheets number of Men to make, and number of Threads each Man, as pay.

	Tacks.				Sheets		
Size.	Number of Men.	Threads each Man.	Pay.	Size.	Number of Men.	Threads each Man.	Pay.
Inches.			qrs. ths.	Inches.			qrs. ths.
3	5	6	5 0	3	9	2	3 0
3½	5	6	5 0	3½	9	3	4 3
4	6	6	6 0	4	9	3	4 3
4½	7	6	7 0	4½	10	3	5 0
5	8	7	9 2	5	12	4	8 0
5½	9	8	12 0	5½	15	4	10 0
6	10	9	15 0	6	19	4	12 4

N.B.—One labourer is allowed to three spinners in the making of all description of cordage. .

LINES AND TWINE.

Deep sea lines	120 fathoms	36	lbs.
Do. do.	120 ,,	32	,,
Do. do.	120 ,,	34	,,
Do. do.	120 ,,	28	,,
Hand lead lines	20 ,,	4	,,
Hambro' lines, 12-thread ..	23 ,,	3	,,
Do. do. 9 ,, ..	23 ,,	2¼	,,
Do. do. 6 ,, ..	23 ,,	1½	,,
Fishing lines	25 ,,	1	,,
Do. do.	25 ,,	¾	,,
Do. do.	25 ,,	½	,,
Do. do.	25 ,,	¼	,,
Samson lines	30 ,,	1½	,,
Do. do.	30 ,,	1¼	,,
Do. do.	30 ,,	1	,,
Do. do.	30 ,,	¾	,,
Log lines, 12-thread	25 ,,	3	,,
Do. 12 ,, ...	25 ,,	1	,,
Houseline and marline	12 skeins	4	,,
Sewing twine	24 ,,	{8 / 9}	,, / ,,
Roping do.	24 ,,	{8 / 9}	,, / ,,

The spinners are paid by the number of threads they spin 160 fathoms, and the number of threads allowed for any rope they make, 6 threads being termed a quarter; thus, if a man spin 40 threads, it is equal 6 quarters and 4 threads, and if he makes two 3-inch ropes, and three 2-inch ropes, there are equal 2 quarters and 2 threads, added together, make 9 quarters at pence per quarter.

Wheel-turners, tenders, and hatchellors are labourers, and are paid 1 penny per quarter less than the spinner, and

One man turns to 12 spinners.
,, ,, tends to 12 ,,
Two men hatchel to 12 ,,
Three ,, top to 12 ,,

Whale lines are dressed by the dozen.

SPINNING MACHINES.

To find the draught of the preparing machine, count the number of teeth in the wheel on the end of feed-roller shaft, and call it the first leader; the number of teeth in the pinion it drives, and call it the first follower; and in like manner all the leaders and followers to the last follower upon the drawing-roller shaft, omitting all intermediate wheels.

Then products of leaders and diameter of roller, divided by product of followers and feed-roller, gives the draught.

SPINNING FRAMES.

Diameter of driving-pulley to speed, so is diameter of driven to the number of revolutions per minute.

CANS AND TUBES.

Diameters of leaders to speed, so is diameter of followers to revolutions.

THE

METHOD PRACTISED IN THE GOVERNMENT ROPEYARDS.

PETERSBURGH HEMP.

The hemp is weighed from the hemp-house to the hatchellor in bundles of seventy pounds for twenty-thread yarn, and the hatchellor takes out seven pounds of shorts, and supplies the spinner with sixty-three pounds of keckled hemp to spin eighteen threads, each one hundred and seventy fathoms, and weigh three and a-half pounds.

EXAMPLE,

Shorts	4·1271
Waste	1·4285

And 70 = 5·5556 = 64·4444 lbs.

Then,

To Spinner	64·4444
Flyings	1·4444

And 63·   ÷ 18 = 3·5 lbs. each thread.

RIGA HEMP.

The bundle weighed to the hatchellor for twenty-five thread yarn is fifty-six pounds, and to the spinner fifty-one pounds, to spin eighteen threads one hundred and seventy fathoms ; weight two pounds eight tenths.

EXAMPLE.

Shorts	3·3585	
Waste	1·6415	
56	5·0000	51 lbs.

Then,

| Spinner | 51· |
| Flyings | ·6 |

$$50·4 \div 18 = 2·8 \text{ lbs.}$$

HEMP ITALIAN.

To spin forty-thread yarn, you take sixteen pounds of shorts from one hundred and twelve pounds of hemp, or about fourteen and a quarter per cent.

The bundles of hemp are divided into hands for the different size yarns or threads.

EXAMPLE.

20 threads	18 hands	3·5 lbs. each.
25 ,,	18 ,,	2·8 ,,
30 ,,	35 ,,	2· ,,
40 ,,	40 ,,	1¾ ,,

RULE.

To find the weight of a thread when size and length are given.

EXAMPLE.

Divide length by size and multiply by seven, and divide the
quotient by seventeen.

EXAMPLE.

170 fathoms of 20-thread yarn.

$$170 \div 20 = 8\cdot5 \times 7 = 59\cdot5 \div 17 = 3\cdot5 \text{ lbs.}$$

Add one sixth for tar, will give the weight of 170 fathoms
tarred.

EXAMPLE.

170 fathoms of 20-thread.

$$3\cdot5 \div 5 = \cdot7 + 3\cdot5 = 4\cdot2 \text{ lbs. tarred.}$$

RULE.

To find the weight of any size yarn 170 fathoms, white.

EXAMPLE.

$$70 \div 20 = 3\cdot5 \quad \text{lb., 20-thread.}$$
$$70 \div 25 = 2\cdot8 \quad \text{,, } 25 \quad \text{,,}$$
$$70 \div 30 = 2\cdot33 \text{ ,, } 30 \quad \text{,,}$$
$$70 \div 40 = 1\cdot75 \text{ ,, } 40 \quad \text{,,}$$

70 being the formula of white yarn, 84 will be the formula
for tarred yarn.

EXAMPLE.

$$70 \div 5 = 14 + 70 = 84 \text{ formula.}$$

Then,

Weight of 170 fathoms of tarred yarn.

$$84 \div 20 = 4\cdot2 \quad \text{lb., 20-thread.}$$
$$84 \div 25 = 3\cdot36 \text{ ,, } 25 \quad \text{,,}$$
$$84 \div 30 = 2\cdot8 \quad \text{,, } 30 \quad \text{,,}$$
$$84 \div 40 = 2\cdot1 \quad \text{,, } 40 \quad \text{,,}$$

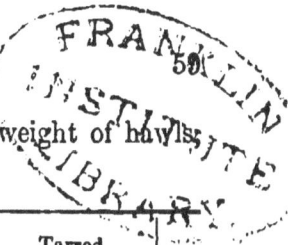

The size of yarn, number of threads, and weight of hawls,
170 fathoms, white and tarred.

				White.			Tarred.		
20 threads 400	..	..		12	2	0	15	0	0
25 ,, 500	..	..		12	2	0	15	0	0
30 ,, 600	..	..		12	2	0	15	0	0
40 ,, 800	..	..		12	2	0	15	0	0

Length to register the lessoms or strands for the different
descriptions of cordage.

EXAMPLE.

Faths. rope.   Faths. strd.

  101 Cable   registered 152

  100 Shroud, 4-strand ,, ... 152

  113 Hawser, 3-strand ,, ... 152

  106   ,,   4-strand ,, ... 152

  122 Bolt-rope   ,, ... 152

    20 4-strand cable stay ,, ... 152

All registered strands, 152 fathoms, are equal 170 warped.
Deduct 21 part, the remainder is the number of threads of
170 fathoms expended in making the rope.

EXAMPLE.

A 6-inch 3-strand hawser contains 240 threads of 20-thread
yarn,

Then,

240 ÷ 21 = 11·4 ... 240 − 11·4 = 228·6 threads = 170
fathoms each.

RULE.

To find the length to register the lessoms or strands for any
length of cables or hawsers.

EXAMPLE.

3-strand cable, 80 fathoms.

$80 \times 152 \div 101 = 120\cdot39$ fathoms.

4-strand cable, 60 fathoms.

$60 \times 152 \div 100 = 91\cdot2$ fathoms.

3-strand hawser, 80 fathoms.

$80 \times 152 \div 113 = 107\cdot6$ fathoms.

4-strand hawser, 60 fathoms.

$60 \times 152 \div 106 = 86$ fathoms.

BOLT-ROPE.

$90 \times 152 \div 122 = 112\cdot13$ fathoms.

TABLES.

Register, hard and laying distance, Three-strand Cables.

	Fathoms.
Length to register	152
Hard distance	10
Hard mark	142
Laying distance	24
Length of strand	118
Cable, hard	2
Hard mark	116
Laying distance	15
Length of cable, when made	101

Four-strand Cable Shrouds.

		Fathoms.
Strand register mark	.. ..	152
,, hard distance	.. ..	10
,, hard mark ..	.. ..	142
,, laying distance ..	.. ..	26
,, length when made	.. ..	116
Cable hard distance..	.. ..	2
,, hard mark ..	.. ..	114
,, laying distance	.. ..	14
,, length when made	.. ..	100

Three-strand Hawser.

		Fathoms.
Register mark ..	.. ..	152
Hard distance ..	.. ..	10
Hard mark	.. ..	142
Laying distance	.. ..	29
Length of rope when made	.. ..	113

Four-strand Hawser.

		Fathoms.
Register mark ..	.. ..	152
Hard distance ..	.. ..	10
Hard mark	.. ..	142
Laying distance	.. ..	36
Length of rope when made	.. ..	106

Stay Strands.

		Fathoms.
Register mark		152
Hard distance		10
Hard mark		142
Laying distance		26
Length of strand made		116

These proportions are 24 fathoms of stay; or, 6·33 fathoms of lessom make 1 fathom of stay.

Bolt-rope.

		Fathoms.
Register mark		152
Hard distance		8
Hard mark		144
Laying distance		22
Length of rope when made		122

Press or Weight.

The rule generally is, one barrel of 5 cwt. to every six threads the strand contains of twenty-thread yarn; and the strands in all cases registered at an angle of 27 degrees, and as much press used while hardening as the strands will bear without injury; and the twist or angle to be brought to 37 degrees, before closing the strands, and to be kept thereat while laying.

PRESS.

A 6-inch Common Hawser.

	Barrels.	Cwt.
At commencement	12 ..	60
Reduce at 5 fathoms	2 ..	10
,, 7½ ,, ...	1 ..	5
Press when hard	9 ..	45
Reduce when started	2 ..	10
And lay the rope with	7 ..	35

6-inch Shroud Hawser.

	Barrels.	Cwt.
Begin with	13 ..	65
Reduce at 5 fathoms	1 ..	5
,, 7½ ,, ...	1 ..	5
Press when hard	11 ..	55
Reduce when started	2 ..	10
And lay the rope with	9 ..	45

12-inch Cable Strand.

	Barrels.	Cwt.
Begin with	12 ..	60
Reduce at 5 fathoms	2 ..	10
,, 7½ ,, ...	2 ..	10
Press when hard	8 ..	40
Reduce when started	2 ..	10
Lay the strand with	6 ..	30

12-*inch Cable*.

	Barrels.	Cwt.
Begin with	32 ..	160
Reduce at 5 fathoms	2 ..	10
„ 2 „	2 ..	10
When hard	28 ..	140
Reduce when started	2 ..	10
Lay the cable with	26 ..	130

12-*inch Stay Strand*.

	Barrels.	Cwt.
Begin with	12 ..	60
Reduce at 5 fathoms	2 ..	10
„ 7½ „	2 ..	10
Press when hard	8 ..	40
Reduce when started	2 ..	10
Lay the strand with	6 ..	30

These are rules laid down for the guidance of the layer, but there are causes that should be noticed as acting upon the strands, such as the atmosphere; it not only has effect upon the strands, but also upon the ground.

RULES.

To find the number of threads per hook or strand for three-strand hawsers.

As the number of threads in the strand of a 10-inch rope, so is any other size squared to the number of threads required.

EXAMPLE.

$$20\text{-thread}—10^2 : 216 :: 6^2 : 77 \text{ threads.}$$
$$25 \text{ ,, } \qquad 10^2 : 272 :: 8^2 : 174 \text{ ,,}$$
$$30 \text{ ,, } \qquad 10^2 : 334 :: 12^2 : 432 \text{ ,,}$$
$$40 \text{ ,, } \qquad 10^2 : 432 :: 5^2 : 108 \text{ ,,}$$

From $4\frac{1}{2}$ inches down, it is as 3^2 to 20.

EXAMPLE.

$$20\text{-thread}—3^2 : 20 :: 4^2 : 36 \text{ threads.}$$
$$25 \text{ ,, } \qquad 3^2 : 25 :: 2^2 : 11 \text{ ,,}$$

Seven-seventeenths the size of the hawsers is the size of the strand.

Eight-fifteenths the size of the cable is the size of the strand ; four-fifteenths size of the cable size of lessom.

RULE.

Four-strand Hawser.

Seven-tenths the number of threads in a three-strand hawser is the number of threads per hook for a four-strand hawser.

EXAMPLE.

Six-inch hawser, four-strand the yarn, twenty-fives.

Then,

$$\text{As } 10^2 : 272 :: 6^2 : 98, \text{ three-strand.}$$
$$98 \times 7 = 68\cdot6 = 69, \text{ four ,,}$$
$$69 \div 3 = 23\cdot = 23, \text{ heart.}$$
$$69 \times 4 = 276 + 23 = 299, \text{ threads in rope.}$$

N.B.—The length of yarn for the heart is one-fourth longer than the rope.

F

To find the number of threads per hook or lessom for three-strand cables.

EXAMPLE.

As 10^2 : 58 :: any other size upwards.

6^2 : 21 :: 　　　„　　　　„　　to nine and a half.

EXAMPLE.

Twelve and eight-inch Cables.

10^2 : 58 :: 12^2 : 84 threads per hook.

6^2 : 21 :: 8^2 : 36　　　„

Four-strand Stays.

Seven-tenths the number of threads in the lessom of three-strand cables is the number of threads for four-strands.

EXAMPLE.

Twelve-inch Stay.

10^2 : 58 :: 12^2 : 84 threads three-strand.

$84 \times 7 = 58 \times 13 = 754$　„　in stay.

Or,

10^2 : 58 :: 8^2 : $37 \times 7 = 25\cdot9 = 26$ threads per hook.

RULE.

Description of hemp and size of yarn for the various rope made.

EXAMPLE.

Cables and cablets, Petersburgh	..	..	20-thread.
Hawsers and shrouds,	Riga	..	..	..	25 ditto.
Bolt-rope, $3\frac{1}{4}$ upwards	ditto	..	..	..	30 ditto.
Ditto　3 downwards	ditto	..	..	..	40 ditto.
Breeching	ditto	..	..	..	25 ditto.
White ropes	ditto	..	..	..	25 ditto.

N.B.—The breechings are laid left-handed or contra way. Bolt-rope and breechings are made from Italian hemp.

WAGES AT H.M. ROPEYARDS.

QUANTITIES OF THE DIFFERENT SORTS OF PRODUCE.

	cwts.	qrs.	lbs.
Hemp	22	2	14
Bands	0	2	14
	22	0	0
Tyres	0	0	24
	21	3	4
Waste	0	0	8
	21	2	24
Flyings and shorts	1	2	24
For yarn	20	0	0

Eight per cent. taken out for common rope; fourteen per cent. taken out for fine rope.

Hemp prepared for one ton of Rope.

	cwts.	qrs.	lbs.
Hemp prepared	16	2	18
Tar	3	1	10
Twenty-thread yarn, 533 threads	20	0	0

Average of yarn spun annually is about 960 hauls at one yard.

Cost of preparing Hemp for spinning Petersburgh.

	cwts. qrs. lbs.					s.	d.
Preparing for 533 threads	..	..	16	2	18	..	18	9½
Do. do. 640 do.	..	..	20	0	0	..	23	1

Italian and Hungarian twenty-five-thread Yarn.

	cwts. qrs. lbs.
Preparing for 687 threads	16 2 18
Do. do. 820 do.	20 0 0

Landing Hemp.

Twenty-five tons one day's work.

					£	s.	d.
30 spinners at 3s. 6d. each	..	..	..	£5	5	0
9 labourers at 2s. each ..	..	..	..	0	18	0

$$£6 \quad 3 \quad 0$$

$123 \div 25 = 4s. 11d. \frac{1}{25}$ per ton.

Parting Hemp.

Seven farthings per bundle.

$7 : 70 :: 2240 : 224 = 4s. 8d.$ per ton.

Carrying Hemp.

$\frac{11}{12}$, or 9·166 part of one penny.

$70 : 9166 :: 2240 : 294 = 2s. 5½d.$ per ton.

Hatchelling.

5½ bundles, or three hundred and eighty-five pounds, for two shillings.

385 : 24 :: 2240 : 11s. 7½d. per ton.

Spinning.

26 threads, or ninety-one pounds, for three shillings and sixpence.

91 : 42 :: 2240 : 86s. 1¾d. per ton.

Wheel Heaving.

208 threads, or seven hundred and twenty-eight pounds, for two shillings.

728 : 24 :: 2240 : 6s. 1¾d. per ton.

Wheel Tending.

208 threads, or seven hundred and twenty-eight pounds, for three shillings and sixpence.

728 : 42 :: 2240 : 10s. 9¼d. per ton.

Boys.

208 threads, or seven hundred and twenty-eight pounds, for two shillings.

728 : 24 :: 2240 : 6s. 1¾d. per ton.

It requires 2488·8 pounds of hemp to produce one ton of white yarn—twenty-thread, or about eleven per cent. more hemp than yarn.

Cost per ton Twenty-thread.

	£	s.	d.
Landing hemp	0	4	11
Carrying do.	0	2	5½
Parting do.	0	4	8
Hatchelling do.	0	11	7½
Spinning do.	4	6	1¾
Wheel-heaving do.	0	6	1¾
Wheel-tending do.	0	10	9¼
Boys do. do.	0	6	1¾
Per ton	£6	12	10½

Salary of Officers.

	£	s.	d.
Master ropemakers, per annum	250	0	0
Foremen do.	200	0	0
Layer do.	150	0	0
Leading man of spinners	70	4	0
Do. do. hatchellors	70	4	0
Do. do. hemp-house	70	4	0
Do. do. hard	70	4	0

The last four are paid 4s. 6d. per day.

RULE.

To find the pay of spinners and labourers making cables, strands, hawsers, stays, &c.

EXAMPLE.

Multiply number of spinners required by 14, and the number of labourers by 11, add together the product, and as the sum is to 14 or 11, so is the pay to each.

A 3-inch hawser requires five spinners and three labourers, and the sum for making is 3s. 10¾d.

Then,

$$5 \times 14 = 70, \text{ and } 3 \times 11 = 33 + 70 = 103.$$
$$103 : 14 :: 3/10\tfrac{3}{4}d. : 6\text{·}354, \text{ spinners' pay.}$$
$$103 : 11 :: 3/10\tfrac{3}{4}d. : 4\text{·}992, \text{ labourers' pay.}$$

MEN AND THEIR OFFICES.

Thirty-nine men land twenty-five tons of hemp.

One man parts one ton of hemp.

One man carries to ten hatchellors.

One man hatchels five and a-half bundles.

One man turns the wheel to eight spinners.

One man tends the wheel to eight spinners.

Two boys tend to eight spinners.

One man attends six tar-kettles.

One man spins twenty-six threads, one hundred and seventy fathoms.

One man superintends forty-eight spinners.

Two labourers warp six hundred and forty threads per day.

Two boys assist labourers warping.

Two spinners setting up yarn.

Two spinners wind five hundred threads.

One boy attends four winding-machines.

TWENTY AND TWENTY-FIVE THREAD YARN.

TABLE showing weight of 1 Thread, 170 Fathoms to 500, White and Tarred.

No.	20-Thread.						25-Thread.					
	White.			Tarred.			White.			Tarred.		
	cwt.	qrs.	lbs.	cwt.	qrs.	lbs.	cwt.	qrs.	lbs.	cwt.	qrs.	lbs.
1	0	0	3½	0	0	4·2	0	0	2·8	0	0	3·36
2	0	0	7	0	0	8·4	0	0	5·6	0	0	6·72
3	0	0	10½	0	0	12·6	0	0	8·4	0	0	10·08
4	0	0	14	0	0	16·8	0	0	11·2	0	0	13·44
5	0	0	17½	0	0	21·0	0	0	14·0	0	0	16·8
6	0	0	21	0	0	25·2	0	0	16·8	0	0	20·16
7	0	0	24½	0	1	1·4	0	0	19·6	0	0	23·52
8	0	1	0	0	1	5·6	0	0	22·4	0	0	26·88
9	0	1	3½	0	1	9·8	0	0	25·2	0	1	2·24
10	0	1	7	0	1	14·0	0	1	—	0	1	5·6
20	0	2	14	0	3	—	0	2	—	0	2	11·2
30	0	3	21	1	0	14·0	0	3	—	0	3	16·8
40	1	1	0	1	2	—	1	0	—	1	0	22·4
50	1	2	7	1	3	14·0	1	1	—	1	2	—
100	3	0	14	3	3	—	2	2	—	3	0	—
200	6	1	0	7	2	—	5	0	—	6	0	—
300	9	1	14	11	1	—	7	2	—	9	0	—
400	12	2	0	15	0	—	10	0	—	12	0	—
500	0	0	0	0	0	—	12	2	—	15	0	—

THIRTY AND FORTY THREAD YARN.

TABLE showing weight of Thread, 170 Fathoms to 800,
White and Tarred.

No.	30-Thread.						40-Thread.					
	White.			Tarred.			White.			Tarred.		
	cwt.	qrs.	lbs.	cwt.	qrs.	lbs.	cwt.	qrs.	lbs.	cwt.	qrs.	lbs.
1	0	0	2·33	0	0	2·8	0	0	1·75	0	0	2·1
2	0	0	4·66	0	0	5·6	0	0	3·5	0	0	4·2
3	0	0	6·99	0	0	8·4	0	0	5·25	0	0	6·3
4	0	0	9·32	0	0	11·2	0	0	7·0	0	0	8·4
5	0	0	11·65	0	0	14·0	0	0	8·75	0	0	10·6
6	0	0	13·98	0	0	16·8	0	0	10·5	0	0	12·6
7	0	0	16·31	0	0	19·6	0	0	12·25	0	0	14·7
8	0	0	18·64	0	0	22·4	0	0	14·0	0	0	16·8
9	0	0	20·97	0	0	25·2	0	0	15·75	0	0	19·0
10	0	0	23·3	0	1	—	0	0	17·5	0	0	21·0
20	0	1	18·6	0	2	—	0	1	7·0	0	1	14·0
30	0	2	13·9	0	3	—	0	1	24·5	0	2	7·0
40	0	3	9·2	1	0	—	0	2	14·0	0	3	—
50	1	0	4·5	1	1	—	0	3	3·5	0	3	21·0
100	2	0	9·0	2	2	—	1	2	7·0	1	3	14·0
200	4	0	18·0	5	0	—	3	0	14·0	3	3	—
300	6	0	27·0	7	2	—	4	2	21·0	5	2	14·0
400	8	1	8·0	10	0	—	6	1	—	7	2	—
500	10	2	17·0	12	2	—	7	2	21·0	9	1	—
600	12	2	—	15	0	—	9	1	14·0	11	1	—
700	0	0	—	0	0	—	10	3	21·0	13	0	14·0
800	0	0	—	0	0	—	12	2	—	15	0	—

HAWSERS TWENTY THREAD YARN.

TABLE showing number of Threads in Strand and Rope, and weight of 113 Fathoms, White and Tarred.

Size.	Threads in Strand.	Threads in Rope.	White.				Tarred.			
Inches.			Ton.	cwt.	qrs.	lbs.	Ton.	cwt.	qrs.	lbs.
1	3	9	0	0	1	2	0	0	1	8
1½	5	15	0	0	1	22	0	0	2	4
2	9	27	0	0	3	6	0	0	3	24
2½	14	42	0	1	1	0	0	1	2	0
3	20	60	0	1	3	4	0	2	0	16
3½	26	78	0	2	1	8	0	2	3	4
4	35	105	0	3	0	14	0	3	3	0
4½	43	129	0	3	3	20	0	4	2	24
5	53	159	0	4	3	8	0	5	3	4
5½	66	198	0	5	3	16	0	7	0	8
6	77	231	0	6	3	14	0	8	1	0
6½	91	273	0	8	0	14	0	9	3	0
7	105	315	0	9	1	14	0	11	1	0
7½	122	366	0	10	3	16	0	13	0	8
8	138	414	0	12	1	8	0	14	3	4
8½	156	468	0	13	3	20	0	16	2	24
9	174	522	0	15	2	4	0	18	2	16
9½	195	585	0	17	1	18	1	0	3	16
10	216	648	0	19	1	4	1	3	0	16
10½	238	714	1	1	1	0	1	5	2	0
11	262	786	1	3	1	16	1	8	0	8
11½	285	855	1	5	1	21	1	10	2	3
12	308	924	1	7	2	0	1	13	0	0

HAWSERS TWENTY-FIVE THREAD YARN.

TABLE showing number of threads in Strand and Rope, and weight of 113 Fathoms, White and Tarred.

Size.	Threads in Strand.	Threads in Rope.	White.				Tarred.			
Inches.			Ton.	cwt.	qrs.	lbs.	Ton.	cwt.	qrs.	lbs.
1½	6	18	0	0	1	20	0	0	2	6
2	11	33	0	0	3	4	0	0	3	12
2½	17	51	0	1	0	24	0	1	1	13
3	25	75	0	1	3	4	0	2	0	16
3½	34	102	0	2	1	20	0	2	3	18
4	44	132	0	3	0	16	0	3	3	2
4½	56	168	0	4	0	0	0	4	3	2
5	68	204	0	4	3	12	0	5	3	9
5½	83	249	0	5	3	20	0	7	0	13
6	98	294	0	7	0	0	0	8	1	17
6½	115	345	0	8	2	24	0	9	3	12
7	134	402	0	9	2	8	0	11	1	26
7½	153	459	0	10	3	20	0	13	0	13
8	174	522	0	12	1	20	0	14	3	17
8½	197	591	0	14	0	8	0	16	3	15
9	220	660	0	15	2	24	0	18	3	12
9½	246	738	0	17	2	8	1	1	0	10
10	272	816	0	19	2	12	1	3	1	7
10½	299	897	1	1	1	12	1	5	2	14
11	329	987	1	3	2	1	1	8	0	22
11½	360	1080	1	5	2	25	1	10	3	24
12	392	1176	1	8	0	0	1	13	2	11

CABLES—TWENTY THREAD YARN.

TABLE showing number of Threads in Lessom and Cable, and
weight of 101 Fathoms, White and Tarred.

Size.	Threads in Lessom.	Threads in Cable.	White.				Tarred.			
Inches.			Ton.	cwt.	qrs.	lbs.	Ton.	cwt.	qrs.	lbs.
2	2	18	0	0	2	4	0	0	2	16
2½	3	27	0	0	3	7	0	0	3	26
3	5	45	0	1	1	10	0	1	2	12
3½	7	63	0	1	3	14	0	2	1	0
4	9	81	0	2	1	17	0	2	3	15
4½	12	108	0	3	0	24	0	3	3	12
5	15	135	0	4	0	14	0	4	3	10
5½	18	162	0	4	3	7	0	5	3	2
6	21	189	0	5	2	14	0	6	3	0
6½	24	216	0	6	1	20	0	7	2	14
7	28	252	0	7	2	0	0	9	0	0
7½	32	288	0	8	2	7	0	10	1	4
8	36	324	0	9	2	16	0	11	2	8
8½	44	396	0	11	3	4	0	14	0	16
9	48	432	0	12	3	10	0	15	1	20
9½	52	468	0	13	3	21	0	16	2	24
10	58	522	0	15	2	4	0	18	2	16
10½	64	576	0	17	0	16	1	0	2	8
11	68	612	0	18	0	24	1	1	3	12
11½	76	684	1	0	1	14	1	4	1	22
12	84	756	1	2	2	0	1	7	0	0
12½	92	828	1	4	2	14	1	9	2	10
13	96	864	1	5	3	1	1	10	3	18
13½	104	936	1	7	3	12	1	13	1	20
14	112	1,008	1	10	0	0	1	16	0	0
14½	120	1,080	1	12	0	13	1	18	2	4
15	128	1,152	1	14	1	4	2	1	0	16
15½	136	1,224	1	16	1	21	2	3	2	24
16	148	1,332	1	19	2	17	2	7	2	8
16½	156	1,404	2	1	3	4	2	10	0	16

CABLES—*continued.*

Size.	Threads in Lessom.	Threads in Cable.	White.				Tarred.			
Inches.			Ton.	cwt.	qrs.	lbs.	Ton.	cwt.	qrs.	lbs.
17	164	1,476	2	3	3	18	2	12	2	12
17½	176	1,584	2	5	0	16	2	16	2	8
18	184	1,656	2	9	1	3	2	19	0	15
18½	196	1,764	2	12	2	0	3	3	0	0
19	208	1,872	2	15	2	24	3	6	3	12
19½	220	1,980	2	18	3	21	3	10	3	25
20	232	2,088	3	2	0	17	3	14	2	8
20½	240	2,160	3	4	1	4	3	17	0	16
21	252	2,268	3	7	2	0	4	1	0	3
21½	264	2,376	3	10	2	24	4	4	3	12
22	280	2,520	3	15	0	0	4	10	0	0
22½	292	2,628	3	18	0	22	4	13	3	12
23	304	2,736	4	1	1	2	4	17	2	24
23½	320	2,880	4	5	2	24	5	2	3	2
24	332	2,988	4	8	3	18	5	6	2	23
24½	345	3,105	4	12	1	7	5	10	3	3
25	360	3,240	4	16	1	17	5	15	2	21

THREE AND FOUR-STRAND HAWSERS.

TABLE showing Size, Number of Men to Make, and Cost of Labour.

Size.	Three-Strand.					Four-Strand.				
	Spinning.	Labour.	Cost.			Spinning.	Labour.	Cost.		
In.			£	s.	d.			£	s.	d.
1½	4	2	0	1	4¾	4	2	0	1	7½
2	5	2	0	1	9¼	5	2	0	2	1
2½	5	2	0	2	3½	5	2	0	2	6½
3	5	3	0	2	10¼	6	3	0	3	3
3½	7	3	0	3	7½	8	3	0	4	6¼
4	9	4	0	5	3¾	10	5	0	6	11½

HAWSERS—*continued.*

Size.	Three-Strand.			Four-Strand.		
	Spinning.	Labour.	Cost.	Spinning.	Labour.	Cost.
In.			£ s. d.			£ s. d.
4½	10	5	0 6 11½	11	6	0 7 10¼
5	11	5	0 8 8½	13	6	0 10 4
5½	12	6	0 9 9	14	7	0 11 4½
6	14	7	0 11 4½	16	8	0 13 0
6½	16	8	0 13 0	18	9	0 17 6½
7	19	9	0 15 2½	21	10	1 0 2½
7½	22	11	0 17 10½	24	12	1 3 4¾
8	24	12	1 3 4½	28	14	1 7 3½
8½	28	14	1 7 3½	31	16	1 10 6
9	31	16	1 10 6	35	18	1 14 4¾
9½	34	17	1 13 1½	39	20	1 18 3½
10	38	19	1 17 0½	43	21	2 12 0¾
10½	41	21	2 0 3	46	23	2 16 0¼
11	45	22	2 3 7	49	25	3 0 0¾
11½	48	24	2 6 9½	53	26	3 4 3
12	51	26	2 10 0	56	28	3 8 3

BOLT-ROPE.

TABLE showing Size of Rope, number of Men to Make, and Cost of Labour.

Size.	Bolt-rope.			Size.	Bolt-rope.		
	Spinning.	Labour.	Cost.		Spinning.	Labour.	Cost.
In.			£ s. d.	In.			£ s. d.
1	3	2	0 1 2½	3	6	3	0 4 2
1¼	3	2	0 1 4	3¼	6	3	0 4 2
1½	4	2	0 1 9¼	3½	7	3	0 4 8
1¾	5	2	0 2 1	3¾	7	4	0 5 1
2	5	2	0 2 3½	4	9	4	0 6 1
2¼	5	2	0 2 3½	4¼	9	4	0 6 1
2½	5	2	0 2 10½	4½	9	5	0 6 5½
2¾	5	3	0 3 2½	4¾	9	5	0 6 5½

Bolt-rope—*continued.*

Size.	Bolt-rope.			Size.	Bolt-rope.		
	Spinning.	Labour.	Cost.		Spinning.	Labour.	Cost.
In.			£ s. d.	In.			£ s. d.
5	10	5	0 8 1½	7	17	8	0 16 3½
5¼	10	5	0 8 1½	7¼	18	9	0 17 6½
5½	11	5	0 8 8½	7½	19	10	0 18 9½
5¾	12	6	0 9 9	7¾	21	10	1 0 2½
6	13	6	0 10 4	8	22	11	1 1 5½
6¼	13	7	0 10 9½	8¼	23	12	1 2 8½
6½	14	7	0 11 4½	8½	24	12	1 3 4¾
6¾	15	8	0 14 10¾				

CABLE STRANDS.

TABLE showing Men to Make, and Cost of Labour.

Size.	Spinners.	Labourers	Cost of Labour.	Size.	Spinners.	Labourers	Cost of Labour.
In.			£ s. d.	In.			£ s. d.
3	4	2	0 2 5	11½	13	6	0 13 2¾
3½	4	2	0 2 5	12	13	7	0 13 11
4	4	2	0 2 9½	12½	15	7	0 15 3¾
4½	4	2	0 2 9½	13	16	8	0 19 5¾
5	5	2	0 4 0¼	13½	17	9	1 1 1½
5½	5	2	0 4 0¼	14	18	9	1 1 11
6	6	2	0 4 11	14½	20	10	1 3 6¾
6½	6	2	0 4 11	15	21	10	1 5 2
7	7	3	0 6 1¾	15½	22	11	1 6 9¾
7½	7	3	0 6 1¾	16	23	12	1 8 5½
8	7	4	0 6 7¾	16½	24	12	1 9 5
8½	8	4	0 7 3¾	17	25	13	1 10 10¾
9	9	4	0 7 11½	17½	27	13	1 12 6
9½	9	5	0 8 5¾	18	28	14	1 14 1¼
10	10	5	0 9 1¾	18½	29	14	1 14 11¼
10½	11	5	0 9 9	19	30	15	1 16 6½
11	12	6	0 12 6½	19½	31	15	1 17 4¼

CABLE STRANDS—*continued.*

Size.	Spinners.	Labourers	Cost of Labour.			Size.	Spinners.	Labourers	Cost of Labour.		
In.			£	s.	d.	In.			£	s.	d.
20	33	16	1	19	9¼	22½	43	21	3	2	4¼
20½	34	17	2	1	4¾	23	45	22	3	5	3¼
21	37	18	2	4	7¼	23½	47	23	3	8	2½
21½	38	19	2	6	3	24	49	24	3	11	2
22	41	20	2	19	5½						

CLOSING CABLES.

TABLE showing Cost of Labour.

Size.	Spinners.	Labourers	Cost of Labour.			Size.	Spinners.	Labourers	Cost of Labour.		
In.			£	s.	d.	In.			£.	s.	d.
3	5	3	0	3	3	14	45	23	2	1	5½
3½	6	3	0	3	7¾	14½	48	24	2	3	10
4	7	4	0	4	5½	15	52	26	2	7	6¼
4½	8	4	0	4	10¼	15½	55	27	2	9	11¼
5	9	4	0	6	4½	16	59	29	3	11	5¾
5½	9	5	0	6	10¼	16½	61	30	3	13	9¼
6	11	5	0	7	10¾	17	65	32	3	18	7¼
6½	12	6	0	8	9¾	17½	69	34	4	3	7½
7	13	7	0	9	9	18	73	36	4	8	5¾
7½	15	7	0	10	9	18½	77	38	4	13	4¾
8	16	8	0	11	8½	19	82	41	5	19	10½
8½	17	9	0	12	8½	19½	86	43	6	5	8
9	19	9	0	13	8	20	90	45	6	11	6½
9½	20	10	0	14	7½	20½	94	47	6	17	4¾
10	23	11	1	0	8¼	21	99	50	7	5	2
10½	24	12	1	1	11	21½	106	53	7	14	11¼
11	27	14	1	4	11½	22	113	56	8	4	7¾
11½	29	15	1	6	9¾	22½	119	60	8	14	5
12	32	16	1	9	3	23	126	63	9	4	1¾
12½	35	17	1	11	8	23½	133	66	9	13	10¾
13	39	19	1	15	3¾	24	139	70	10	3	7¼
13½	41	21	1	17	9						

STAY STRANDS.

TABLE showing number of Men and Cost of Labour.

Size.	Spinners.	Labourers	Cost of Labour.			Size.	Spinners.	Labourers	Cost of Labour.		
In.			£	s.	d.	In.			£	s.	d.
7	7	3	0	6	1¼	14	13	7	0	13	11
7½	7	3	0	6	1¼	14½	14	7	0	17	0½
8	7	4	0	6	3¾	15	15	7	0	17	10½
8½	8	4	0	7	4	15½	16	8	0	19	5¾
9	8	4	0	7	4	16	17	9	1	1	1½
9½	9	4	0	7	11¼	16½	18	9	1	1	11
10	9	4	0	7	11¼	17	19	10	1	3	6½
10½	9	5	0	9	9	17½	20	10	1	4	4
11	10	5	0	10	5¼	18	21	11	1	6	0½
11½	10	5	0	10	5¼	18½	23	11	1	7	7¾
12	11	5	0	11	1¾	19	24	12	1	9	3
12½	11	5	0	11	1¾	19½	25	12	1	10	0¾
13	12	6	0	12	6½	20	26	13	1	11	8
13½	13	6	0	13	2¾						

CLOSING STAYS.

TABLE showing number of Men and Cost of Labour.

Size.	Spinners.	Labourers	Cost of Labour.			Size.	Spinners.	Labourers	Cost of Labour.		
In.			£	s.	d.	In.			£	s.	d.
7	15	7	0	5	5	14	32	16	1	3	4¾
7½	15	7	0	5	5	14½	32	16	1	3	4¾
8	17	8	0	8	1¾	15	35	17	1	5	3¾
8½	17	8	0	8	1¾	15½	35	17	1	5	3¾
9	19	10	0	9	5	16	37	19	1	7	3¾
9½	19	10	0	9	5	16½	37	19	1	7	3¾
10	21	11	0	10	5	17	40	20	1	9	3
10½	21	11	0	10	5	17½	40	20	1	9	3
11	24	12	0	11	8½	18	43	22	1	11	8
11½	24	12	0	11	8½	18½	47	23	1	14	1¼
12	27	13	0	19	5¾	19	50	25	1	16	6½
12½	27	13	0	19	5¾	19½	53	27	1	18	11¼
13	29	15	1	1	5¾	20	57	28	2	1	4¼
13½	29	15	1	1	5¾						

STRANDS, TRANSPORTING CABLES.

TABLES showing number of Men and Cost of Labour.

Size.	Spinners.	Labourers	Cost of Labour.	Size.	Spinners.	Labourers	Cost of Labour.
In.			£ s. d.	In.			£ s. d.
5	5	2	0 4 10¼	7½	7	4	0 8 11¾
5½	5	2	0 4 10¼	8	7	4	0 8 11¾
6	5	3	0 5 7	8½	8	4	0 9 9
6½	5	3	0 6 6½	9	9	4	0 10 7
7	7	3	0 8 1¾				

CLOSING TRANSPORTING CABLES.

Size.	Spinners.	Labourers	Cost of Labour,	Size.	Spinners.	Labourers	Cost of Labour.
In.			£ s. d.	In.			£ s. d.
5	9	4	0 7 0¾	7½	15	7	0 13 4¾
5½	9	5	0 7 7½	8	16	8	0 16 8½
6	11	5	0 8 8½	8½	17	9	0 18 1
6½	12	6	0 10 11¾	9	19	9	0 19 5¾
7	13	7	0 12 3¾				

TACK STRANDS.

TABLES showing number of Men and Cost of Labour.

Size.	Spinners.	Labourers	Cost of Labour.			Size.	Spinners.	Labourers	Cost of Labour.		
In.			£	s.	d.	In.			£	s.	d.
3	3	2	0	0	$8\frac{1}{4}$	7	4	2	0	0	$11\frac{3}{4}$
$3\frac{1}{2}$	3	2	0	0	$8\frac{1}{4}$	$7\frac{1}{2}$	4	2	0	0	$11\frac{3}{4}$
4	3	2	0	0	$8\frac{1}{4}$	8	5	2	0	1	5
$4\frac{1}{2}$	3	2	0	0	$8\frac{1}{4}$	$8\frac{1}{2}$	5	2	0	1	5
5	3	2	0	0	$9\frac{1}{4}$	9	5	3	0	1	5
$5\frac{1}{2}$	3	2	0	0	$9\frac{1}{4}$	$9\frac{1}{2}$	5	3	0	1	$7\frac{1}{2}$
6	4	2	0	0	$11\frac{3}{4}$	10	6	3	0	1	$10\frac{1}{4}$
$6\frac{1}{2}$	4	2	0	0	$11\frac{3}{4}$						

CLOSING TACKS.

Size.	Spinners.	Labourers	Cost of Labour.			Size.	Spinners.	Labourers	Cost of Labour.		
In.			£	s.	d.	In.			£	s.	d.
3	5	3	0	2	2	7	10	5	0	6	$1\frac{1}{4}$
$3\frac{1}{2}$	5	3	0	2	2	$7\frac{1}{2}$	11	5	0	8	$8\frac{1}{4}$
4	6	3	0	2	5	8	11	6	0	9	$2\frac{3}{4}$
$4\frac{1}{2}$	7	3	0	3	3	$8\frac{1}{2}$	12	6	0	9	9
5	7	4	0	3	7	9	13	6	0	10	$3\frac{1}{2}$
$5\frac{1}{2}$	8	4	0	3	$10\frac{3}{4}$	$9\frac{1}{2}$	13	7	0	10	10
6	9	5	0	5	$8\frac{3}{4}$	10	14	7	0	11	5
$6\frac{1}{2}$	9	5	0	5	$8\frac{3}{4}$						

EQUALIZING MACHINE.

There are so many of these machines in use of various makes and forms, that to lay down a rule for the one would not do for another; the only method to be obtained is to give one turn or twist to the strand or readie, while the machine draws it a certain length; a common jack could be made to perform this work by drawing it a certain distance while it made a certain number of revolutions: thus, the driving wheel 18 inches, the pinions 3 inches, will drive the pinions six revolutions to one; therefore, in making a strand for a 6-inch rope, it requires to have one turn or twist in 6 inches of length, then while the driving wheel makes one revolution, the jack or machine should be 36 inches or 3 feet, and in all cases the same.

The Equalizing or Forming Machine is so constructed, that if the whelps of the barrel that carries the ground-rope be set 6 inches from the centre of the shaft, it will be the set for the strand of a 6-inch rope, and the set for all other strands are to set the whelps so that it will draw a certain length to one twist.

EXAMPLE.

The machine making six to one.

3-inch rope, set the whelps 3 inches.
4-inch	„	„	„	4 inches.
5-inch	„	„	„	5 inches.
6-inch	„	„	„	6 inches.

Some make a small difference for strands or bolt-rope to suit their purpose, but it is not requisite.

TABLE showing how to set the machine for making Cordage upon a length.

	Below.	Before.
Three-strand hawsers	10	to 6½
Four-strand ditto	10	„ 7
Cable-strand	4	„ 3
Cables—three-strand	10	„ 7
Bolt-rope	7½	„ 6
Cables—four-strand	5½	„ 4

It must be understood that to make a rope it requires more fore-twist than after ; the strand requires three-tenths of one turn more than the rope in three-strand hawsers, therefore the extra seven-tenths given the rope is to overcome friction of top, tails, supports, &c. Then, if we give the rope two twists to one, and three-tenths in the strand, it is as 10 to 6½.

Three-strand Hawsers, 113 *and* 142.

Then,

As Length of strand	..	..	142	..	..	1·15229
: Radius	..	.. ..	90	..	..	10·
:: Length of rope	..	..	113	..	..	·05308
: Cosine >	..	..	37° 17′	..	..	9·90079

Therefore,

As Tangent	..	..	..	37° 17′	..	..	9·88158
: Radius	..	..	.. ..	90	..	..	10·
:: Circumference	..	..	..	6 in.	..	1·77815
: Length of strand	..		7·88	..	..	1·89657

And,

As Radius	..	..	..	..	90	..	.. 10·
: Circumference	..	..	..	6 in.		..	1·77815
:: Tangent	..	..	..	37° 17′	..	..	9·88158
: Length of strand		..	7·88	..		..	1·89657

Then, as 6 : 1 : : 78·8 : 1·3 ; or, as 6 inches to one turn in the rope, so is 1·3 to one turn in the strand ; then, two in the rope to 1·3 in the strand equal 10 to 6½.

●

Cable Strand, 118 *and* 142.

Then,

As Strand at hard	..	..	142	..	..	1·15229
: Radius	..	..	..	90	..	.. 10·
:: Strand made	..	..	.. 118	..	..	1·07188
: Cosine >	..	..	..	33° 48′	..	..	9·91959

As Tangent	..	..	33° 48′	..	..	9·82571
: Radius	..	..	..	90	..	.. 10·
:: Circumference	..	..	6 in.		..	1·77815
: Length of strand		..	8·96	..	..	1·95244

And,

As Radius	..	..	..	90	..	.. 10·
: Circumference	..	..	..	6 in.	..	1·77815
:: Tangent	..	..	..	33° 48′	..	..	9·82571
: Length of strand		..	8·96	..	..	1·95244

Then say 9;

And as 6 : 1 : : 9 : 1·5, or one and a half; or as 6 inches to one turn in strand, so is 1½ to one turn in ready; and say 2 in strand to 1·5; it is equal 3 or 4, or 4 turns below to 3 before, in making cable strands.

Bolt-rope, 144 *to* 122.

Then,

As Length of strand	144	.. ..	1·159868
: Radius	90	.. ..	10·
:: Length of bolt-rope ..	122	.. ..	1·088136
: Cosine >	32° 2′	.. ..	9·928268

Therefore,

As Tangent	32° 2′	.. ..	9·796351
: Radius	90	.. ..	10·
:: Circumference	6 in.	..	1·778151
: Length of strand ..	9·59	.. ..	1·981800

And,

As Radius	90	.. ..	10·
: Circumference	6 in.	..	1·778151
:: Tangent	32° 2′	.. ..	9·796351
: Length of strand ..	9·59	.. ..	1·981800

Then as 6 : 1 : : 9·6 : 1·6 turns in strand; and 2 below to 1·6 before is equal to 7½ below to 6 before.

THE SLIDE RULE MADE EASY.

MULTIPLICATION.

Rule :—

A |	Multiplicand.	Product.
B |	1	Multiplier.

EXAMPLE I.

Multiply 15 by 4; referring to the form, and placing figures for the case, it will stand thus :—

A |	15	60
B |	1	4

Place 1 upon B under 15 upon A, and over 4 on B is the product 60 on A.

EXAMPLE II.

Multiply 8·5 by 6·5.

A |	8·5	55·25
B |	1	6·5

EXAMPLE III.

Multiply 9 by 20.

A |	9	180
B |	1	20

DIVISION.

Rule :—

A	Quotient.	Dividend.
B	1	Divisor.

EXAMPLE I.

Under the dividend place the divisor, then over unit will be the answer.

Divide 120 by 5.

A	24	120
B	1	5

EXAMPLE II.

Divide 110 by 8.

A	13·75	110
B	1	8

EXAMPLE. III.

Divide 150 by 6.

A	25	150
B	1	6

DIRECT PROPORTION OR RULE OF THREE.

Rule :—

A	Second term.	Fourth term.
B	First term.	Third term.

EXAMPLE I.

If 4 lbs. cost 9d., what will 20 lbs. cost ?

A	9	45d., or 3s. 9d.
B	4	20

EXAMPLE II.

If 2 give 7, what will 40 yield?

A	7	140
B	2	40

EXAMPLE III.

If 1 cwt. of rope cost 40s., how much for 7 lbs.?

A	25 or 2s. 6d.	40
B	7	112

Place 112 upon B and over it on the line A 40, then over 7 on the line B will be 25 or 2s. 6d. for the 7 lbs.

INVERSE PROPORTION.

Rule :—

A	Second term.	Fourth term.
C	First term.	Third term.

EXAMPLE I.

If a thread of eighteens, 160 fathoms, weigh 4 lbs., what will a thread of twenties weigh, 160 fathoms?

A	18	20
C	4	3·6 lbs., answer.

EXAMPLE. II.

If 24 men spin 50 cwt. in 2 days, how many days will be required for 8 men to spin the same?

A	2	6 answer.
C	24	8

The same may be obtained without reversing the slide.

EXAMPLE III.

A		Second term.	Fourth term.
B		Third term.	First term.

SQUARES AND ROOTS.

The squares and roots are to be found upon the lines C and D, without disturbing the slide.

C	1		Square.
D	1		Root.

EXAMPLE I.

Find the squares of 3, 7, and 9.

C	1	9	49	81 answer squares.
D	1	3	7	9

EXAMPLE II.

Find the squares of 4, 8, and 12.

C	1	16	64	144
D	1	4	8	12

'Unit upon the line C, over unit on the line D, will be squares on the line C and roots on the line D.

MENSURATION.

Circles.

Any circle twice the diameter of another contains twice the circumference and four times the area; therefore the circumferences are as their diameters, and their areas as the squares of their diameters; and as 1 to 3·1416 or 7 to 22, so is the diameter to the circumference.

EXAMPLE I.

The diameter 50 inches, what the circumference?
Rule :—

A	22	157 answer.
B	7	50

EXAMPLE II.

The diameter 20 inches, required the circumference.

A	3·1416	62·8320 answer.
B	1	20

RULES FOR ROPEMAKERS.

To find the weight of a thread 160 fathoms, size eighteens, white.

EXAMPLE I.

A	4	72
B	1	18

EXAMPLE II.

The weight of a thread 160 fathoms, size eighteens, tarred.

A	5	90
B	1	18

EXAMPLE III.

The weight of a thread 160 fathoms, size twenties, white.

A	3·6	72
B	1	20

Example IV.

The weight of a thread 160 fathoms, size twenties, tarred.

A	4·5 or 4½ lb.	90
B	1	20

LENGTH OF YARN.

Three-strand Hawser.

Example I.

130 fathoms of rope required.

A	3	195 answer.
B	2	130

Four-strand Hawser.

Example II.

90 fathoms of rope required.

A	11	141 answer.
B	7	90

Three-strand Cable.

Example III.

120 fathoms of cable required.

A	5	200
B	3	120

Four-strand Cable.

Example IV.

30 fathoms of cable required.

A	7	52
B	4	30

THREADS PER HOOK.

Three-strand Hawsers.

EXAMPLE I.

6-inch yarn, twenties.

C | 2   80 answer.
      ———————————————————————
D | 3   6

Four-strand Hawser.

EXAMPLE II.

6-inch yarn, twenty-fives.

C | 25   67 answer.
      ———————————————————————
D | 3·68   6

Three-strand Cables.

EXAMPLE III.

12-inch cable, yarn eighteens.

C | 18   72
      ———————————————————————
D | 6   12

Four-strand Cables.

EXAMPLE IV.

9-inch cable, yarn eighteens.

C | 18   30
      ———————————————————————
D | 7   9

LENGTH TO FORM THE STRAND.

Three-strand Hawsers.

EXAMPLE I.

130 fathoms of rope required.

A | 7   182 answer.
      ———————————————————————
B | 5   130

Four-strand Hawser.

EXAMPLE II.

100 fathoms of rope required.

A		7¼		145 answer.
B		5		100

Three-strand Cables.

EXAMPLE III.

120 fathoms of cable required.

A		3		180 answer.
B		2		120

Four-strand Cables.

EXAMPLE IV.

40 fathoms of cable required.

A		11		63
B		7		40

TUBES' DIAMETERS.

Three-strand Hawsers.

EXAMPLE I.

9-inch rope required.

A		·5		1·5 or 1½ diameter.
B		3		8

Four-strand Hawsers.

EXAMPLE II.

8-inch rope required.

A		10		1·14 diameter.
B		7		9

Three-strand Cables.

EXAMPLE III

20-inch cable required.

A	10	1·66 diameter.
B	12	20

Four-strand Cables.

EXAMPLE IV.

14-inch cable required.

A	·5	1 diameter.
B	7	14

LONDON:
Printed by W. CLOWES & SONS, Stamford Street and Charing Cross.